Jesus Monteiro

Fluid Mechanics and Turbomachinery

Jesus Monteiro

Fluid Mechanics and Turbomachinery

Development of an aerodynamic tunnel for testing Turbomachinery

ScienciaScripts

Imprint

Any brand names and product names mentioned in this book are subject to trademark, brand or patent protection and are trademarks or registered trademarks of their respective holders. The use of brand names, product names, common names, trade names, product descriptions etc. even without a particular marking in this work is in no way to be construed to mean that such names may be regarded as unrestricted in respect of trademark and brand protection legislation and could thus be used by anyone.

Cover image: www.ingimage.com

This book is a translation from the original published under ISBN 978-620-4-19218-5.

Publisher:
Sciencia Scripts
is a trademark of
Dodo Books Indian Ocean Ltd., member of the OmniScriptum S.R.L Publishing group
str. A.Russo 15, of. 61, Chisinau-2068, Republic of Moldova Europe
Printed at: see last page
ISBN: 978-620-4-07611-9

Acknowledgements

I thank in the first instance Jehovah (God), for his protection during days and nights. Then, to the Faculty of Engineering of the Agostinho Neto University of Angola, in the person of Mr. Professor Valter Lourenço de Jesus for having believed in me.

I would also like to express my gratitude to Professor Jorge Marques Ramalheira, for the knowledge he passed on to me during my training as a Mechanical Engineer at this University.

I also thank Mr. Professor António Carlos Mendes, my scientific advisor at the University of Beira Interior (UBI), for the support provided during the realization of this work.

I thank Mr. António Morgado, Technical Assistant of the Fluid Mechanics and Turbomachinery Laboratory of UBI, for his support in the experimental setup. To Francisco Pavão Braga, researcher at the Laboratory, I thank for all the support received and the information provided. To Bernardo Gomes, Laboratory colleague, I thank for all the support received and the information provided. To Mr. José Catalão, a locksmith from UBI's Workshop, I thank the help provided in the design and fabrication of some elements of the Tunnel Aerodynamics. The University of Beira Interior, for the conditions that were made available to me to carry out this research.

Finally, I thank my family for their unconditional support during the completion of this work.

Summary

Wind tunnels have often been used to validate the results of mathematical models applied to the studies of turbomachinery blade crowns. Therefore, the main objective of this work is to develop a wind tunnel for testing turbomachinery, operating in the laboratory of Fluid Mechanics and Turbomachinery at the University of Beira Interior. In a first phase, it was necessary to carry out a literature review on wind tunnels developed for testing turbomachinery worldwide. During the assembly of our tunnel were used some elements already designed and built since 2011. These elements were connected to the propulsion system and are supported on a structure that we designed in Solidworks® 2013 and built in the general workshops of UBI.

The Tunnel was subsequently tested and the results obtained were compared with the predictions of the computational model made with Fluent® 2016. The work developed culminated in a functional installation with great potential for the future development of gas turbines in the Laboratory.

Keywords

Wind Tunnels, Turbomachinery, Axial Fans, Experimental Tests, Computer Simulation.

General Index

List of symbols

Symbol	Meaning	SI
A	Area	[m2]
D	Diameter	[m]
d_{ie}	Inlet internal diameter	[m]
d_{is}	Internal diameter of output	[m]
G	Power generation	m2/s2
g	Gravity acceleration	m/s2
Δh	Height difference	[m]
I	Electric current intensity	[A]
k	Kinetic energy of turbulence	m2/s2
K	Adiabatic constant of air	-
l	ComPeting	[m]
$\dot{m}$	Mass flow rate	kg/s
p_1	Process Pressure	[Pa]
p_2	Reference pressure	[Pa]
p_0	Stagnation pressure	[Pa]
p_{∞}	Static pressure	[Pa]
P_{atm}	Atmospheric pressure	[Pa]
P_e	Inlet pressure	[Pa]
p'	Pressure Correction	[Pa]
p^*	Assumed pressure	[Pa]
P_h	Hydraulic power	[W]
Δp	Pressure difference	[Pa]
Q_i	Elementary flow rate	m3/s
Q_t	Total flow rate	m3/s
q	Quantity of movement	kgm/s
R	Universal constants of ideal gases	J/kgK

R_e	Reynolds Number	-
r	Radius	[m]
S	Fonte da property	kg/ms3
Skn	Skewness	-
T	Local temperature	[K]
T_{∞}	Optimal cell size	[m]
T_o	Cell size	[m]
t	Time	[s]
U	Electrical voltage	[V]
u	Speed component on abscissa axis	m/s
V	Volume	[m3]
$\vec{V}$	Speed vector	[m/s]
v	Flow velocity	[m/s]
$V_{mi,i+1}$	Average speed between two consecutive points	[m/s]
v	Velocity component on ordinate axis	[m/s]
w	Velocity component in the coordinate axis	[m/s]
Y_M	Floating expansion	kg/ms3
β	Inclination of the pressure gauge	[°]
δ	Distance between two end nodes	[m]
ϵ	Rate of dissipation	m2/s3
η_h	Hydraulic performance	[%]
θ_e	Cell inclination	[°]
μ	Absolute viscosity	[Pas]
μ_t	Turbulent viscosity	[Pas]
ν	Kinematic viscosity	[m2/s]
ρ_{ar}	Density of air	[kg/m3]
ρ_{H_2O}	Density of water	[kg/m3]
φ	Property Carry-Forward Function	-
$\cos\varphi$	Power factor	-
Γ_{φ}	Diffusion coefficient	m2/s

Chapter 1

Introduction

The first wind tunnels were an important tool to leverage the development of aircraft, which later led to the conquest of airspace. Currently, in addition to their use in the aeronautical sector, wind tunnels have been used in various areas of industry, particularly in the automotive sector, in civil engineering and architecture, and in the environmental and energy sector. In all these applications, wind tunnels serve to model the interactions of an air flow with the physical model of the facility under study.

The use of super-computers has contributed greatly to a rapid growth in the aerodynamic field and areas mentioned above, allowing experimental results to interact with the results of computer simulations as they progress. They also allow sharing of simulation results in near real time. With the development of today's supercomputers it is already possible to design complex machines through numerical simulation. These simulations allow a quick and effective consideration of all the variables involved in the project, a task that would be complex and costly to perform only through wind tunnel tests. However, the predictions of the mathematical models that serve as a basis for computer simulation invariably require meticulous and accurate validation by means of a physical model tested in the tunnel. The capability of these super-computers has improved at a substantial rate, but not to a sufficient level to replace experimental test facilities for design development.

This work addressed different aspects related to the development of an aerodynamic tunnel for testing Turbomachinery. These aspects include the theoretical foundation of the project, the assembly of the installation, its testing and calibration, the comparison of the measured parameters with the data from the computer simulation and, finally, the main conclusions of the study carried out.

1.1 Objective of this study

The object of the present work is to assemble and test an aerodynamic tunnel that is being developed at the Fluid Mechanics and Turbomachinery Laboratory of UBI *(fluidslab)* [1]. This facility has been developed in the *fluidslab* since 2011, having been designed and built some elements of the Tunnel. Our work consists in assembling the different elements to the propulsion system, which is an Aeric fan, Golden Lebey model, and then manufacturing a support structure of the main body of the Tunnel. This structure was designed in Solidworks® 2013 and built in the University's workshops. Once completed the tests and the Tunnel calibration, the collected measurements are used for comparison with the predictions of the computational simulation performed with the aid of Fluent® 2016.

1.2 Origin and evolution of the first wind tunnels

The first wind tunnels were linked to the aeronautical sector. In the 19th century there were great advances towards the conquest of space. To achieve mastery of the airspace, the aeronautics of the time resorted to the ideas of Leonardo da Vinci and the laws of Isaac Newton. At that time, different models were studied, either stationary or moving through the air. For this

7

purpose were designed and tested various mechanisms, such as the rotating arm.

At this time several European and American aeronauticists were particularly noteworthy. Benjamin Robins (1707-1751), an English engineer and mathematician, was the first to build a 1.219 m long rotary arm (Fig. 1.1).

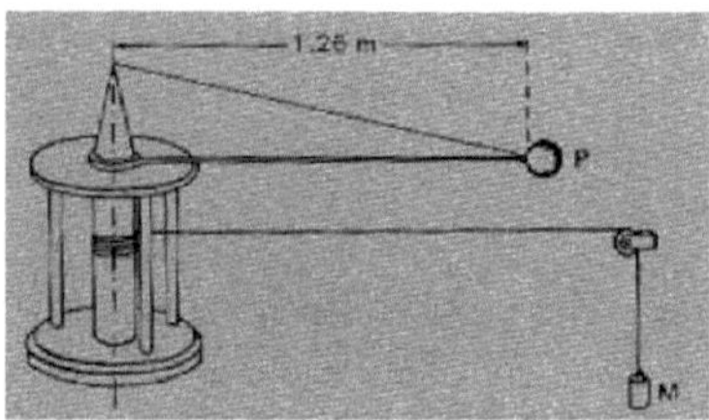

Figure 1.1 - Representation of the Benjamin Robins rotating arm. Source: Ref. A1 (Nasa wind tunnel history)

This arm rotated by means of a descending weight mounted on a pulley. Robins' experiments led to the conclusion that the theories on the resistance to the advance of aerodynamic models, known at the time, were quite inaccurate; models of identical configuration, but with different frontal area, did not always have comparable coefficients of resistance of form [A1].

Sir. George Cayley (1773-1857), who was an English engineer, is world renowned as having been the founder of modern aeronautics. Cayley was the first aeronautical research scientist to publish his work, which includes the principles and accounts for the aerodynamic forces that sustain aircraft flight. George Cayley's rotating arm (Fig. 1.2) was 1.524 m long and rotated with a tangential velocity ranging from 3.048 m/s to 6.096 m/s. The various experiments performed by Cayley revolutionized the way people thought about aeronautics, especially in terms of aircraft design [7].

Figure 1.2 - Representation of George Cayley's rotating arm.

Source: Ref. 7 (A short history of the European transonic wind tunnel ETW, 2011)

Despite providing a lot of aerodynamic information, the use of rotating arms caused some problems. Due to the great turbulence generated by the system, the measurement results suffered considerable deviations from the expected value. In particular, it was not possible to determine the precise value of the relative velocity of the model in relation to the air, and it was difficult to assemble the measuring equipment, especially instruments to measure the small forces exerted on a model rotating at high speed. Limitations of this nature made man to think of a more efficient system.

Thus, the idea arose to use a closed device that allowed the circulation of air along its length, by means of a fan. This device came to be called a wind tunnel. The use of wind tunnels allowed overcoming the problems registered with the use of rotating arms, i.e., with the wind tunnels it became simpler to record the aerodynamic forces during the model tests [A1].

The British Frank H. Wenham (1824-1908), designed and tested in 1871 the first wind tunnel [A1]. Wenham worked a wooden log 3.6576 m long and 0.209 ^{m2} to carry out its assembly. The tunnel was connected to a fan driven by a steam engine, which propelled the air into it. During the experiments, Wenham and his colleagues used bodies of different shapes in order to record the forces of resistance and support generated by the air as a result of its interactions with the objects. During these experiments, important results in the field of aerodynamics were obtained. Wenham's team was able to demonstrate that, for small attack angles, the lift was considerably higher than the resistance.

Subsequently, many other inventors used wind tunnels to test their models. In 1890 the Danish inventor Poul la Cour used a wind tunnel during the development phase of a wind turbine. Carl Rickard, in 1897, also used a wind tunnel to develop his glider called *Flugan*. Also in this year, Konstantin Tsiolkovsky built an open section tunnel with a centrifugal fan to determine the resistance coefficients of flat plates, cylindrical elements and spheres.

The emergence of wind tunnels was seen as a driving factor in understanding the aerodynamics of bodies, in particular the forces of lift and resistance. Nevertheless, Osborne Reynolds (1842-1912) made an important discovery at the University of Manchester, linked to these phenomena. Reynolds demonstrated that the flow of air around model-scale objects was similar to the flow around the full-scale object. Osborne Reynolds defined a dimensional parameter, called Reynolds number, to govern the similarity conditions concerning shape resistance and viscous resistance between model and prototype [A1].

In August 1899 the Wright brothers (Wilbur (1867-1912) and Orville (1871-1948)), built a flying machine in the USA. In this case it was the unmanned biplane kite, 1.524 m long. Later, in 1900, the Wright brothers built a manned aircraft. The major drawback of this aircraft was related to the fact that it generated little lift and much resistance, contrary to what was expected. In order to improve performance the Wright brothers used atmospheric circulation (the wind) to test profiles of different configurations in an open tunnel. The purpose of the experiment was to compare the lift forces for the different geometries [7].

Figure 1.3 - Wind tunnel developed by the Wright brothers in 1901. Source: Ref. A2 (1901 wind tunnel tests)

The first wind tunnel developed by the Wright Brothers had a square section. The results obtained during the tests were so encouraging that the brothers immediately decided to build a larger facility (Fig. 1.3), with $0.1652 \ m^2$.

In summary, until 1920, several wind tunnels were installed in different regions according to table.1.1 below:

Table 1.1: Existing wind tunnels up to 1920 and their location.

Source: Ref. 1 (Contribution to the development of wind tunnel for the testing of turbomachinery blade crowns , master's thesis, University of Beira Interior, Covilhã, 2011)

Dimensions	Date	Location	Owner
0,4x0,4 ^{m2}	1901	Dayton, Ohio, USA	Wright Brothers
1,82x1,82 ^{m2}	1901	Catholic University, USA	Zahm
0.6 m (diameter)	1903	National Physical Laboratory, England	Stanton
1x1 ^{m2}	1903	Rome, Italy	Crocco
1.2 m (diameter)	1904	Koutchino, Moscow, Russia	Riabouchinsky
2x2 ^{m2}	1908	Gottinge, Germany	Prandtl
1.5 m (diameter)	1909	Champ de Mars, France	Eiffel
1,21x1,21 ^{m2}	1910	National Physical Laboratory, England	-
2,13x2,13 ^{m2}	1912	National Physical Laboratory, England	-
2 m (diameter)	1912	Auteuil, France	Eiffel
-	1912	Aachen, Germany	Junkers
2,43x2,43 ^{m2}	1913	Washington Navy Yard, USA	Zahm
1,21x1,21 ^{m2}	1914	MIT, USA	Hunsaker
2,2x2,2 ^{m2}	1916	Gottingen, Germany	Prandtl
1.68 m (diameter)	1917	Stamford University, USA	Durand
2.13 m (diameter)	1918	Hompstead, New York, USA	Curtiss
2,13x4,26 ^{m2}	1918	National Physical Laboratory, England	-
1,37 m (diameter)	1918	Bureau of Standards, USA	-
1,21x1,21 ^{m2}	1919	MIT, USA	Ober
2.29 m (diameter)	1919	Stanford University, USA	Durand

1.3 Dissertation structure

This work is structured into six chapters. Thus, the first chapter describes the objectives of this study, as well as the origin and evolution of wind tunnels until 1920. The second chapter initially presents the history of gas turbines and the experimental facilities that have served for their development until the present day. In the third chapter we focus on the *fluidslab* tunnel. The various constituent elements and the design of a support structure for the main body of the tunnel carried out in Solidworks® 2013 are described. The manufacturing process of this structure in the general workshops of UBI is also described, as well as the procedures for assembling the installation in the Laboratory. The fourth chapter refers to the testing and calibration of the wind tunnel, presenting measured values for the static pressure and velocity distribution in the sections, as well as the turbine flow rate and fan power. The fifth chapter describes the computational model that was carried out with the aid of Fluent® 2016. The results are interpreted here in the light of the measurements performed during the Tunnel test. The results of the computational simulation of the flow in the Tunnel, performed with the aid of Ansys-Fluent® 2016 software, are presented. Finally, in the conclusion of the dissertation, we summarize the main conclusions to be drawn from this study.

Chapter 2

Bibliographic survey

This chapter first presents the history of gas turbines. Then, the experimental facilities for testing Turbomachinery in the United States of America, Canada, Europe and Asia are presented.

2.1 Origin and evolution of gas turbines

Turbomachinery are rotating systems that continuously interact with the fluid by means of a rotor, removing energy from the fluid in the case of a turbine or adding energy to the fluid in the case of a pump or compressor. They are widely used in industry, namely: in the process of power generation, water supply network, etc. Turbomachines are also used as propulsion elements, especially in the aeronautical industry.

Currently there is no consensus about the beginning of the development of Turbomachinery. Some sources refer that the study began about 150 years before Christ (BC) [A3] and others refer 62 years after Christ (AD) [8]. This development was attributed to the hero of Alexandria in the 120s BC, who used a device called the *Aeolipile* that rotated on the action-reaction principle (Fig. 2.1). The device consisted of a bowl with water, heated at the bottom by a boiler. The boiler communicated with a sphere by means of two empty tubes. The sphere has two nozzles which induce its rotation.

Later, in the year 1500 AD, Leonardo Da Vinci described a *chimney* called *"chimney Jack"* (Fig. 2.2). This chimney allowed air to pass around a set of blades, similar to a fan, acting as a turbine and making a vertical axis rotate. The vertical shaft is connected to a horizontal shaft by means of a bevel gear.

Figure 2.1 - *Aeolipile,* developed by the hero of Alexandria (120 BC). Source: Ref. 8 (The Historical Evolution of Turbomachinery)

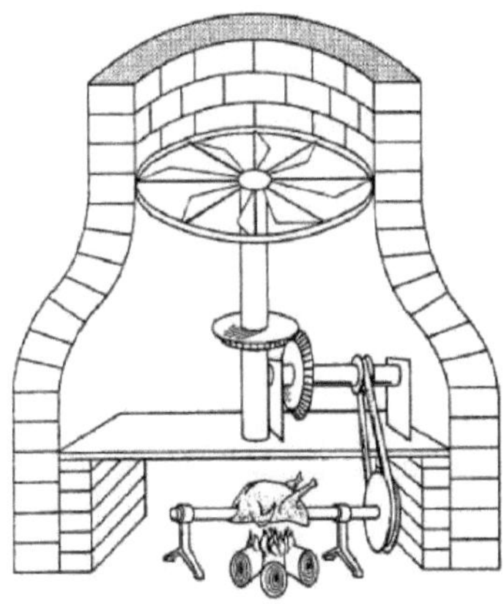

Figure 2.2 - Leonardo Da Vinci's chimney (1500 AD).

Source: Ref. 8 (The Historical Evolution of Turbomachinery)

After the interventions of Leonardo Da Vinci, several ideas were emerging around the development of Turbomachines. For example, in 1629 an Italian engineer named Giovanni Branca designed and built an impulse turbine used to power a stamping mill. Later, in 1687, the formulation of the laws of motion by Sir Isaac Newton catapulted the development around the perception of Turbomachines. Based on these laws a steam powered means of transport consisting of four wheels, a heated spherical boiler and a nozzle designed to provide a reaction jet was developed (Fig. 2.3).

Figure 2.3 - Newton's steam transport.

Source: Ref. 8 (The Historical Evolution of Turbomachinery)

Regarding the origin of gas turbines, although there is no evidence, it is commonly accepted that John Barber patented in 1791 [8], in England, a machine that used the thermodynamic cycle of modern gas turbines (Fig. 2.4). The turbine was basically composed of a reciprocating compressor and a combustion chamber. The uncertainty hovered on which would be the most efficient flammable fuel to move the system. The gas produced would be stored in a common reservoir and then transferred to the combustion chamber where it would be mixed with the air coming from the compressor leading to the explosion. The equipment also had a water cooling system to cool the turbine parts.

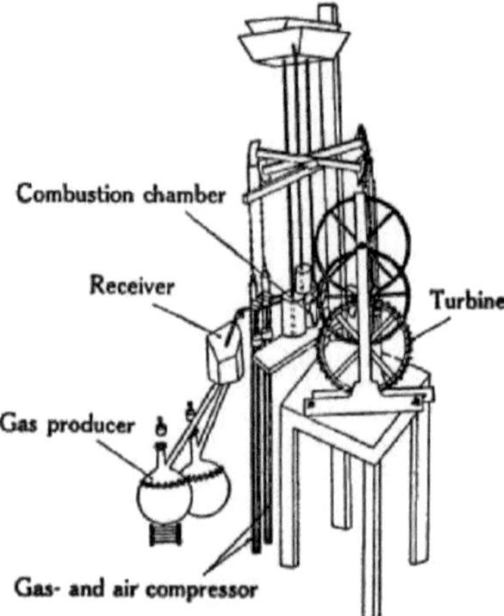

Figure 2.4 - Cycle of a gas turbine, patented by John Barber (1791). Source: Ref. 8 (The Historical Evolution of Turbomachinery)

At the European level, since there is no evidence of the existence of John Barber's patent, the first gas turbine was officially developed in 1872 by Dr. J. Franz Stolze. This German scientist was born in March 1836 in Berlin, and graduated from the University of Berlin in Philosophy, History, Geography, Physics and Mathematics. In 1863 he completed his doctorate on gas turbine design at the University of Jena.

The Stolze turbine was tested between 1900 and 1904 [8]. It was a gas turbine consisting of a 10-stage axial compressor with a pressure ratio of 2.5:1 and a 15-stage reaction turbine. The 25 stages are supported on a single shaft supported at the ends by bearings. The system is connected via belts to an alternator with 150 kW of power. The turbine operated with an inlet temperature around 400 °C.

In the USA the first patent for a gas turbine was filed in the name of Charles G. Curtis in 1895. This inventor is also known for developing the Curtis steam turbine while a researcher at *General Electric* (GE). However, the legend of GE is Stanford Moss, due to his work on supercharged engines in aircraft. Moss graduated from the University of California (UCSF) in 1900, where he researched gas turbine design. S. Moss would later design an experimental gas turbine during his doctoral studies at Cornell University. The turbine drove a compressor that propelled air into the combustion chamber. During the development of this turbine it was found that the energy required by the compressor was greater than the energy delivered and therefore no net work was done.

Independently, in 1905, Dr. Holzworth developed a constant volume gas turbine (Fig. 2.5).

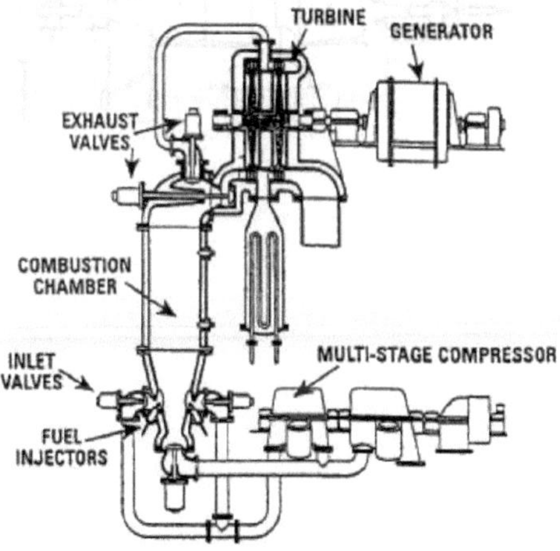

Figure 2.5 - Gas turbine developed by Holzworth (1905). Source: Ref. 8 (The Historical Evolulution of Turbomachinery)

In this turbine the fuel is injected into a closed combustion chamber, where it is mixed with compressed air. The explosion creates a pressure increase 4.5 times higher than the initial value. In this case, the power required by the compressor is a fraction of the power produced by the combustion turbine.

The use of gas turbines for electricity generation started in the 1960s [9]. Since then, this technology has been strongly developed, such that in 30 years each generating unit had a capacity of 100 MW. Currently the design of gas turbines involves some basic or fundamental components, namely: an axial compressor, used to compress the fluid; a combustion chamber or a heat exchanger, to raise the temperature at the output of the mixture; a turbine, with the purpose of producing mechanical work.

Nowadays, the reduction of pollutant emissions is a recurrent concern in the design of electricity generating systems. In this context, computational programs (CFD) have been used to analyze the flow characteristics of alternative fuels such as ethanol, palm methyl ester, etc. The effects of high temperature of the gases resulting from combustion and the fluid dynamic effects are also among the relevant concerns in gas turbine design.

2.2 Test facilities developed in the USA and Canada

This subsection will discuss some wind tunnels that stand out due to their characteristics suitable for the study of Turbomachinery. In the United States of America (USA), most of the wind tunnels with these characteristics were developed by the "National Aeronautics and Space Administration" (NASA), with the exception of a few, present in Universities and private institutions. Among the most important we list:

■ NASA LSAWT wind tunnel

This is a low-speed wind tunnel (Fig. 2.6), developed by NASA's Langley Research Center (LaRC). It was used for testing military and commercial aircraft prototypes. It underwent some modifications in 2016, one of those modifications was the extension of the inlet nozzle. These modifications have expanded the capabilities of the tunnel for aerodynamic and acoustic tests on small turbomachinery rotor systems. This is a free-jet, open wind tunnel, 10.4 m long with a test section measuring 5.2 mx5.2 m. This tunnel generates flows at Mach < 0.32.

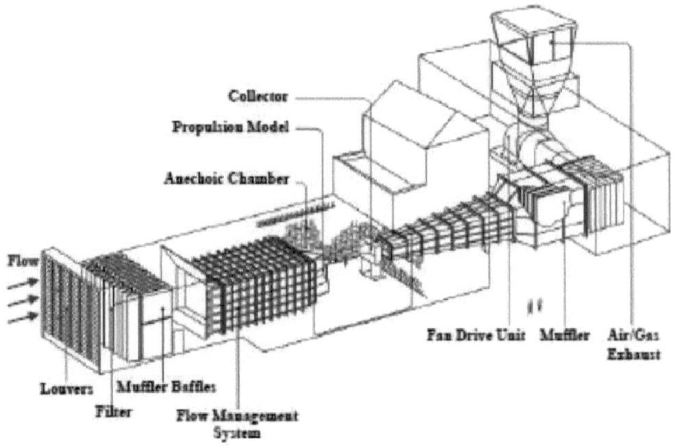

Figure 2.6 - LSAWT wind tunnel, NASA.

Source: Ref. 10 (Small propeller and rotor testing capabilities of the NASA Langley low speed aeroacoustic wind tunnel, 2017)

■ Glenn Research Center, NASA

The Glenn Research Center has a LSWT low-speed wind tunnel (Fig. 2.7), built in 1969, with a cross-section of 2.743 mx3.658 m and Mach number not exceeding 0.23. This device is used to perform aerodynamic and acoustic tests on open rotor blades.

Figure 2.7 - LSAWT wind tunnel, NASA.

Source: Ref. 11 (Initial investigation of the acoustics of a counter-rotating open rotor model with historical baseline blades in a low-speed wind tunnel, 2012)

■ Langley 14x22 ft subsonic tunnel

This is a facility manufactured in 1970, initially used for testing aircraft models in subsonic regime. In 1982 the facility (Fig. 2.8) underwent some modifications, having been resized to house a test section of 4 mx7 m. The facility has the capacity to perform acoustic tests, aircraft rotor tests, among others.

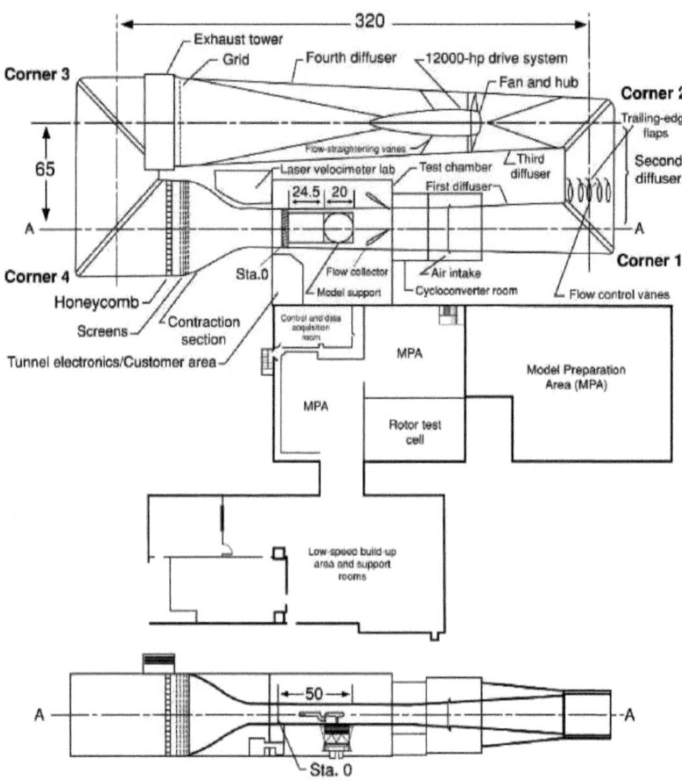

Figure 2.8 - Langley 14x 22 ft subsonic tunnel.

Source: Ref. 12 (Advancing Test Capabilities at NASA Wind Tunnels, 2015)

■ Glenn's 10x10 wind tunnel

This facility (Fig. 2.9) was built in 1955 in order to perform supersonic tests, at high Reynolds number. For this purpose, two compressors with a total power evaluated at 215.321 MW are used. In subsonic conditions, the flow can reach Mach numbers of up to 0.4, while in supersonic conditions Mach numbers can reach 4.1 in certain areas. This tunnel has a square test section with 3.048 m side.

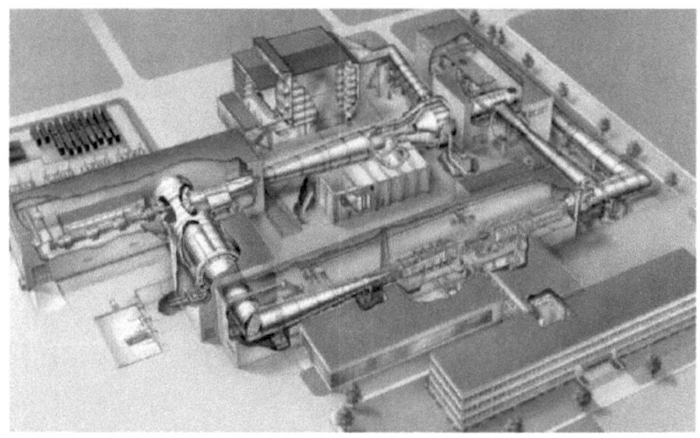

Figure 2.9 - Glenn's 10x10 tunnel.

Source: Ref. 12 (Advancing Test Capabilities at NASA Wind Tunnels, 2015)

The characteristics of this facility enable tests to be carried out on open rotors with little disturbed flow. The low disturbance of the flow makes the tunnel suitable for tests in laminar flow regimes.

■ Glenn's 8x6 and 9x15 wind tunnel

Figure 2.10 shows the 8x6 and 9x15 wind tunnel belonging to the Glenn research center:

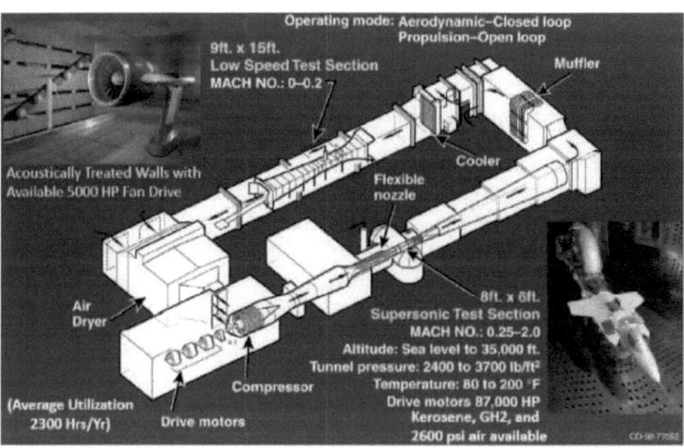

Figure 2.10 - Glenn's 8x6 and 9x15 Tunnel.

Source: Ref. 12 (Advancing Test Capabilities at NASA Wind Tunnels, 2015)

This wind tunnel consists of two test sections, one for supersonic tests and one for subsonic tests. The section used for supersonic tests has a dimension of 2.438 mx1.829 m. In this section the flow develops with a Mach number between 0.25 and 2, pressures between 1,149.12 kPa and 1,771.56 kPa, and temperatures between 26.67 °C °ê 93.33 C. Whereas the section used for the tests in subsonic flow regime has a fan with capacity evaluated in 3,728.5 kW, capable of generating Mach number flows up to 0.2.

This tunnel allows aerodynamic testing of model aircraft and open-circuit propulsion elements.

■ Langley National Transonic Tunnel

The facilities (Fig. 2.11) were manufactured in 1983 to operate with two different gases, namely air and nitrogen. The test section has a dimension of 2.5 mx2.5 mx7.6 m and the pressures vary between 101 and 910 kPa. During its operation with air, the Mach number varies between 0,2 and 1,05 and temperatures vary between 32 °C and 65 C. Under these conditions, the maximum Reynolds number is 65x106/m. This tunnel is used for testing aircraft models and for open-circuit rotors.

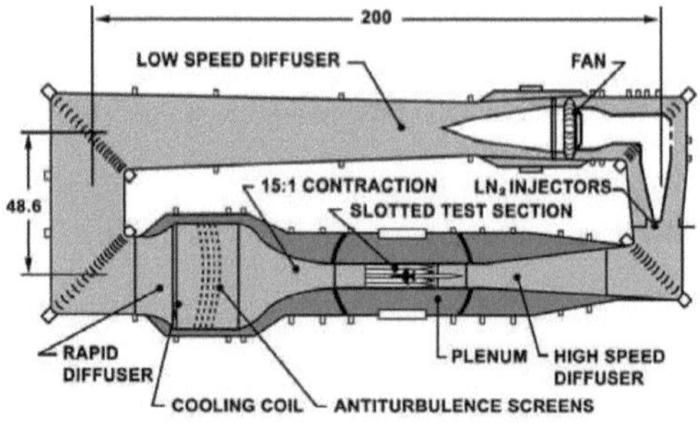

Figure 2.11 - Langley transonic tunnel.

Source: Ref. 12 (Advancing Test Capabilities at NASA Wind Tunnels, 2015)

■ Low Speed Wind Tunnel, University of New Mexico

The Department of Mechanical and Aerospace Engineering at the University of New Mexico has a low-speed wind tunnel with a 4 mx4 m square entrance section. The test section is 1.2 m wide, 14.6 m long and 1.2 m high. The flow velocity in the test region can reach 35 m/s. This wind tunnel was used for testing wind turbine blades. The wind tunnel of the University of New Mexico is shown in figure 2.12.

Figure 2.12 - Subsonic wind tunnel, University of New Mexico.

Source: Ref. 13 (Experimental Study of Turbulence Intensity Influence on Wind Turbine Performance and Recovery in a Low-Speed Wind Tunnel, 2017)

■ University of Pennsylvania Subsonic Wind Tunnel

The Department of Aerospace Engineering at the University of Pennsylvania has a closed-loop subsonic wind tunnel, whose test section is 1.013 m high and 1.476 m wide. In the test section the turbulence intensity is about 0.05%, at a speed of 46m/s [A3]. This tunnel is used for aerodynamic tests on wing profiles, aircraft models and wind turbines.

■ Carleton University Transonic Wind Tunnel

Unlike those of others, the Department of Mechanical and Aerospace Engineering at Carleton University, Ottawa (Canada), has an inclined wind tunnel (Fig. 2.13). The tunnel is used for transonic measurements on cascades of turbine blades in irrotational flow. These measurements make it possible to predict the quality of the flow in contact with the blades and thus improve their performance. The tunnel is coupled to four air tanks with a pressure of 810.4 kPa. When the reservoirs are full, the control valve opens and the air flows inside the tunnel test section, where the turbulence intensity is 0.04%. The blade cascades inside the test section are mounted on a turntable and the angle of incidence of the flow at the blades is frequently adjusted according to the test conditions.

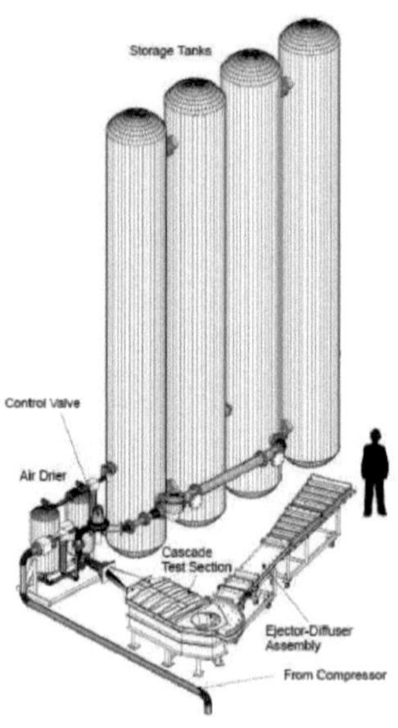

Figure 2.13 - High speed wind tunnel, Carleton University.

Source: Ref. 14 (Impact of Flow Quality in Transonic Cascade Wind Tunnels, 2002)

2.3 Facilities developed in Europe and Asia

According to the report "Aeronautical wind tunnels Europe and Asia" [15], published in February 2006, there are currently a total of 155 wind tunnels, divided into subsonic, supersonic and hypersonic. Table 2.1 indicates the number of existing wind tunnels in several countries in Europe and Asia, as well as their flow regime, including those where the regime is unknown.

Table 2.1: Aeronautical wind tunnels distributed across Europe and Asia.

Source: Ref. 15 (Aeronautical wind tunnels Europe and Asia, 2006).

Location		Subsonic	T ransonic	Supersonic	Hypersonic	Unknown
Europe	Belgium	4	0	2	2	
	England	13	2	3	0	
	France	4	1	4	5	
	Germany	9	3	0	2	
Location		**Subsonic**	**T ransonic**	**Supersonic**	**Hypersonic**	**Unknown**
Europe	Italy	2	1	0	1	
	Netherlands	4	1	5	0	
	Scotland	3	0	0	0	
	Sweden	1	1	0	0	
	SubTotal	40	9	14	10	0
Asia	China	14	5	11	6	4
	Indunesia	1	0	0	0	
	Japan	3	0	8	5	1
	Malaysia	1	0	0	0	
	Singapore	0	1	0	0	
	South Korea	21	0	0	0	
	SubTotal	40	6	19	11	5
	Total	80	15	33	21	5

■ University of Sheffield low speed wind tunnel

Under a project led by Dr. Robert J. Howell, a low-speed wind tunnel for aerodynamic testing was built at the Department of Mechanical Engineering of the University of Sheffield (Fig. 2.14).

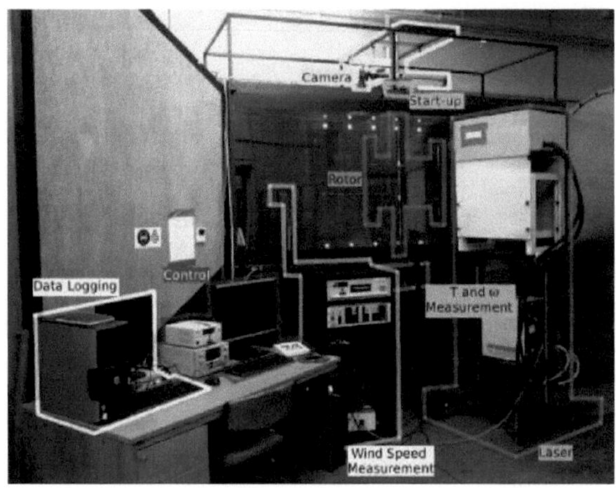

Figure 2.14 - Wind tunnel at the University of Sheffield.
Source: Ref. A5

The maximum velocity achieved by the flow along the tunnel is 22 m/s in a test section of 1.2 mx1.2 mx3 m. The tunnel has a contraction ratio of 6:1 and a turbulence intensity of 0.5%.

- **QinetiQ Facilities, Famborough, England**

QinetiQ is a company formed by scientists and engineers committed to meeting the prevailing needs in the engineering field. In particular, QinetiQ operates mainly in the defence, security and aerospace market. This institution has a low-speed closed-circuit wind tunnel built in 1977, 5 m long and with a test section of 4.2 mx5 mx8 m. The Mach number varies between 0.05 and 0.34, and the maximum values of Reynolds and stagnation pressure are 7.6×10^6 and 3 bar, respectively.

- **Facilities of the University of Liverpool**

In 1966, the University of Liverpool published a paper concerning measurements of turbulence levels in Turbomachinery. The tests were performed in three different conventional subsonic wind tunnels, whose estimated maximum Reynolds number per unit length was 4×10^6. The first tunnel (Fig. 2.15) had a working section of 0.483 mx0.483 m and a maximum flow velocity of 21.336 m/s. This tunnel is coupled to an axial compressor of 9.694 kW of power.

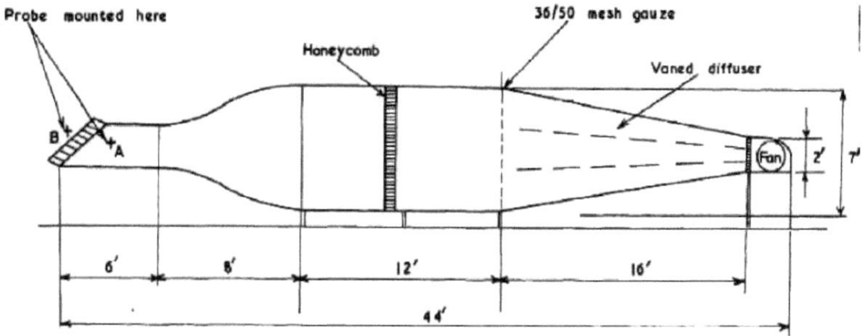

Figure 2.15 - Wind tunnel 1 at the University of Liverpool.

Source: Ref. 16 (Measurement of Turbulence in the Liverpool University Turbomachinery Wind Tunnels and Compressor, 1966)

The second tunnel (Fig. 2.16) had a rectangular test section of 0.483 mx0.305 m, a maximum flow velocity of 21.336 m/s and is coupled to an axial compressor with 11.186 kW of power.

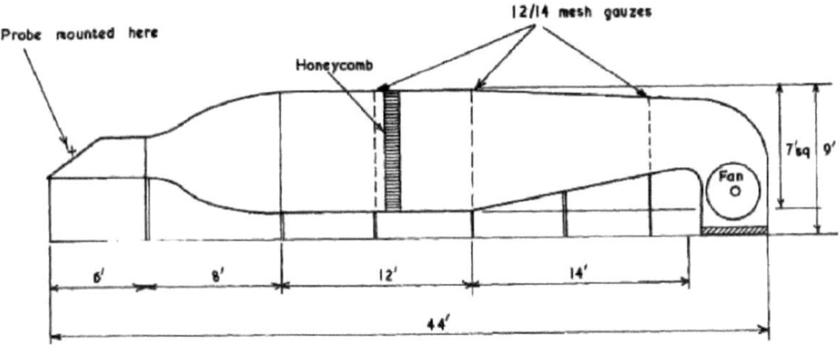

Figure 2.16 - Wind tunnel 2 at the University of Liverpool.

Source: Ref. 16 (Measurement of Turbulence in the Liverpool University Turbomachinery Wind Tunnels and Compressor, 1966)

The third tunnel (Fig. 2.17) of rectangular section (1.219 mx0.609 m) has a maximum flow velocity of 19.81 m/s and is driven by an axial compressor with 14.914 kW of power.

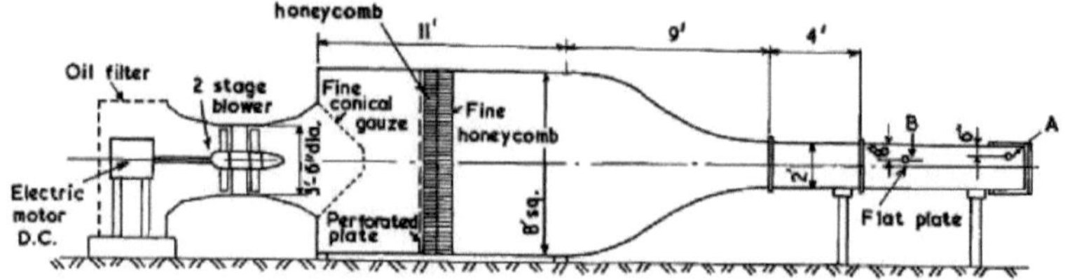

Figure 2.17 - Wind tunnel 3 at the University of Liverpool.

Source: Ref. 16 (Measurement of Turbulence in the Liverpool University Turbomachinery Wind Tunnels and Compressor, 1966)

■ Technical University of Carolo-Wilhelmina, Braunschweig, Germany.

The Institute of Fluid Mechanics at the Technical University of Carolo-Wilhermina, Germany, has a low-speed, low-noise (LNB) wind tunnel (Fig. 2.18) with a test section of 0.6 mx0.4 mx1.5 m, where the flow velocity is 19 m/s. This is a wind tunnel driven by a 3 kW electric motor that is located downstream of the facility. This installation is used for Turbomachinery studies in low Reynolds number flow.

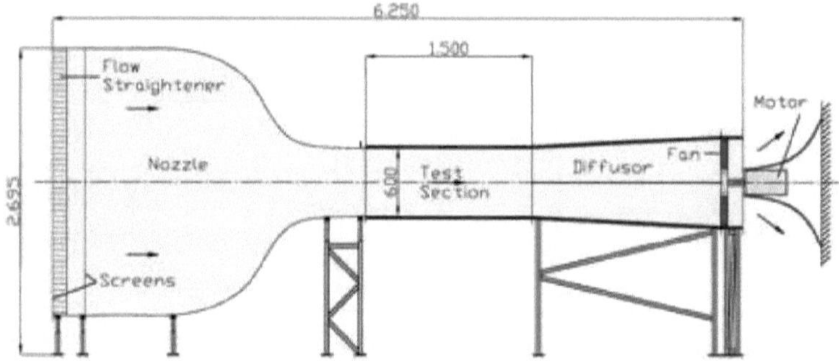

Figure 2.18 - Wind tunnel of the Carolo-Wilhermina LNB.

Source: Ref. 15 (Aeronautical wind tunnels Europe and Asia, 2006)

■ Facilities in Gottingen, Germany

This facility has a high-pressure subsonic wind tunnel (HDG), whose test section has a dimension of 0.6 mx0.6 m, with maximum values of velocity and Reynolds of 35 m/s and 12x106, respectively. Under these conditions, the maximum pressure inside the tunnel is 10,000 kPa. The HDG (Fig. 2.19) is used for several types of investigations, namely for the analysis of wing profiles of aircraft models and turbomachinery blades.

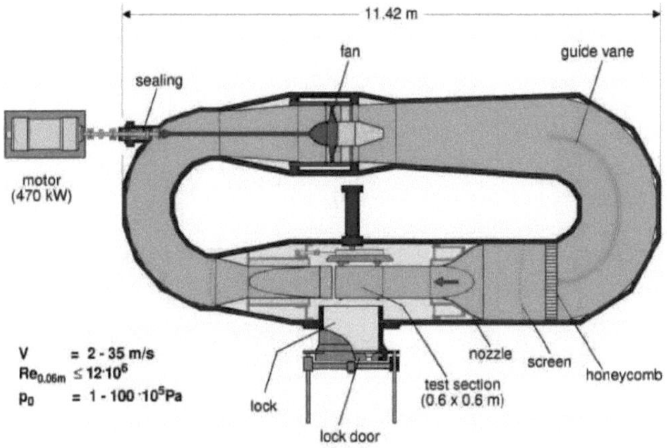

Figure 2.19 - High pressure subsonic wind tunnel (HDG).

Source: Ref. 15 (Aeronautical wind tunnels Europe and Asia, 2006).

- **Wind tunnel L-1 A of the VKI**

The Von Karman Institute (VKI) is an international educational and scientific organization that owns a low-speed wind tunnel, called L-1.

The L-1 is a low-speed wind tunnel with a test section of 3 m diameter and 4.5 m length. Its 4.2 m diameter counter-rotating fans are driven by a DC, 580 kW motor, generating flow with a variable velocity between 2 m/s and 60 m/s. It presents a turbulence intensity of 0.3% and a contraction ratio of 4 [A6]. This wind tunnel was designed to test wing profiles and aircraft models up to 2 m long, in flows with Reynolds up to 4x106. However, its characteristics also allow testing wind turbines.

- **VKI T'-3 wind tunnel**

This is a cryogenic, transonic, pressurised, adjustable-wall wind tunnel built by ONERA as part of a pilot project in Toulouse. The device has a test section 0,1 m wide, 0,6 m long and 0,117 m high, with an internal wall of about 5 mm of cork. This tunnel is currently at the VKI in Belgium.

Initially the tunnel was designed to operate in transonic regime at Mach = 0.8, with a pressure of 4 bar and a temperature of 120 K, obtained through the injection of liquid Nitrogen. Under these conditions the 125 kW fan operates at 11,400 rpm. Currently, at the VKI, the tunnel operates with a 7.5 kW engine and the flow develops at room temperature, with Mach = 0.23.

In addition the test section has been re-dimensioned and is now 800 mm long with a total deflection of 130 mm.

- **Facilities of Instituto Superior Técnico de Lisboa (1ST)**

IST has two wind tunnels for testing Turbomachinery, one horizontal (Fig. 2.20) and one vertical (Fig. 2.21).

The horizontal wind tunnel is driven by a radial fan. While the vertical tunnel operates in a transonic regime and features a variable test section.

The most used test section, the vertical tunnel, has the dimension 30 cmx2 cm. The tunnel is coupled to a compressor with a pressure ratio of 200 kPa.

Figure 2.20 - Horizontal wind tunnel for testing Turbomachinery of 1ST.

Source: Ref. 1 (Contribution to the development of wind tunnel for testing turbomachinery blade crowns, 2011)

Figure 2.21 - Vertical wind tunnel for testing 1ST blade cascades.
Source: Ref. 1 (Contribuição para o desenvolvimento de túnel de vento para o ensaio de coras de blades de Turbomáquinas, 2011)

■ VZLU wind tunnels

VZLU is a company specialized in developing electronic and scientific instruments for space and terrestrial applications, located in the Czech Republic. Within its services the company has several wind tunnels for testing Turbomachinery. The Mach number of the flow in this tunnel varies between 0.1 and 0.98. Under these conditions, Reynolds number varies between 0.06 and 1.2x106.

■ Chalmers University of Technology wind tunnel

This is a tunnel developed under the EU AIDA project in 2004 in Sweden. The facility was modified to meet the needs of another EU project, called AITB-2, in 2005. This tunnel (Fig. 2.22) was used to optimise the performance of gas turbines. It is a semi-open, axisymmetric wind tunnel designed to study turbine heat transfer. The axial velocity at the tunnel entrance is 30 m/s, with a 100 kW power fan.

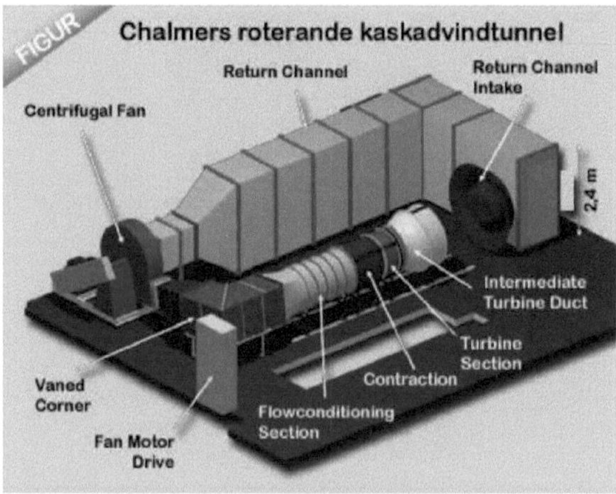

Figure 2.22 - Wind tunnel at low speed.

Source: A9

- **Wind tunnel of Northwest China Polytechnic University, Xian**

This Institute has had an NPU WT52 supersonic wind tunnel since May 1990. The tunnel has a test section of 21.4 cm high, 30 cm wide and 1.08 m long. The facility operates with a Mach number < 1.3 [15]. The tunnel walls, 1 mm thick, are made of a bronze alloy, which is flexible and waterproof. The tunnel has 16 manually operated pressure inlets, for static pressure measurement.

- **Wind tunnel South Korea Defence and Development Agency (ADD)**

This South Korean Agency has a laboratory equipped with a low-speed wind tunnel, built in 1998 at an estimated cost of about US$ 12 million. The tunnel has a test section of 3 mx2.25 mx8.75 m, with flow velocity varying between 10 m/s and 120 m/s. Under these conditions, the stagnation pressure is equivalent to atmospheric pressure, with a maximum Reynolds number of 8x106. The propulsion system of the tunnel consists of an 11-bladed fan with 2,400 kW of power [15].

This facility is used for testing aircraft wing profiles, Turbomachinery blades, missiles, projectiles and water vehicles. During tests with this tunnel, aerodynamic forces, dynamic stability, aero-elasticity, etc. are measured.

- **Wind tunnel of the Technical University of Malaysia**

The Faculty of Engineering at the Technical University of Malaysia has a low-speed wind tunnel (UTM-LST), shown in Figure 2.23.

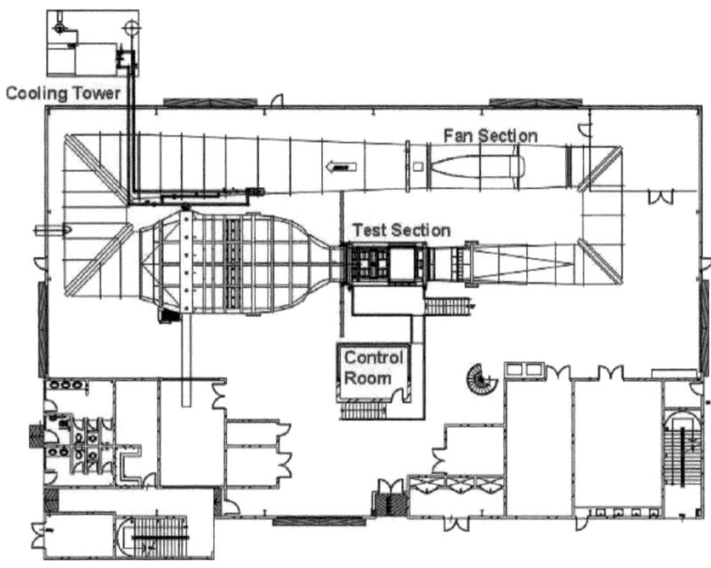

GROUND FLOOR PLAN

Figure 2.23 - Low speed wind tunnel (UTM).

Source: Ref. 15 (Aeronautical wind tunnels Europe and Asia, 2006)

This tunnel, built in May 2001, is at the Aeronautics Laboratory. The costs of its construction are estimated at US$ 7 million. The test section of this tunnel is 1.5 mx2 mx5.8 m, with a flow velocity varying between 3 and 80 m/s. The Reynolds number can reach 1x106 and the stagnation pressure is equivalent to atmospheric pressure. The tunnel is horizontal and has a contraction radius of 9. This Tunnel is used for analysis of aircraft wing profiles in the low speed flow regime, as well as in the analysis of the interaction of air with Turbomachinery blades.

■ Baghdad University wind tunnel

Baghdad University has a closed, low-speed wind tunnel (Fig.2.24). The test section is square, with 0.7 m side. The diffuser element of the tunnel is made of steel, with a length of 6 m, allowing connecting the test section to the propulsion element of the tunnel. The propulsion system consists of an axial alternating current fan, with 55.928 kW power and 3000 rpm rotational speed.

31

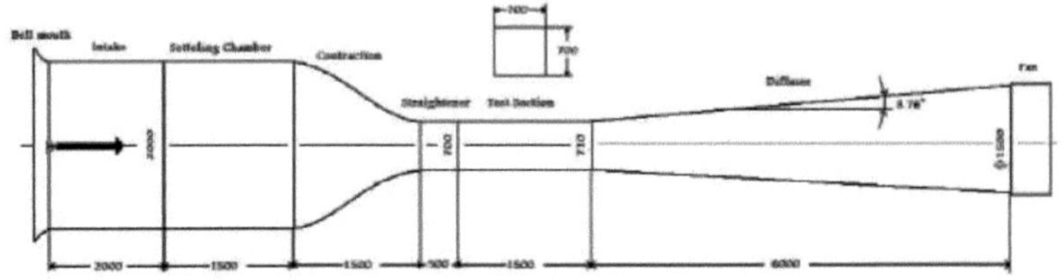

Figure 2.24 - Wind tunnel at low speed.

Source: Ref. 18 (Testing and Commissioning of a Low-Speed Wind Tunnel (LSWT) Test Section, 2014)

Chapter 3

Development and assembly of the *fluidslab* installation

      This chapter presents the constituent elements of the installation, as well as its assembly steps. This is a closed wind tunnel with a non-uniform cross section. The tunnel has a circular cross section along its entire length, with an internal diameter of 600 mm at the entrance and 800 mm at the exit. The total length of the installation is 2,848 m.

3.1 Constituent Elements of the Tunnel

      Table 3.1 shows the elements that make up the main body of the Tunnel, where characteristics such as the internal diameter at the entrance (**die**), internal diameter at the exit (**dis**) and the length of each of the elements are identified:

Table 3.1: Characteristics of the Tunnel elements.

Elements	die [mm]	dis [mm]	l [mm]
Stabilisation stator	600	600	250
Divergent nozzle	600	800	1000
Connection nozzle to the fan	800	800	750
Discharge nozzle	800	800	598

It is important to mention that the elements referenced in Table 3.1 were designed and manufactured within the project named PTDC/EME-MFE/111763/2011-16 *"Design and Experimental Testing of a Power Take-off System for the ANACONDA Wave Energy Conversion Device",* co-financed by the European Union/FEDER [1].

The propulsion system of the Tunnel is composed of an axial fan, which is used to provide the required flow velocity along the Tunnel and to resist the energy losses inside the Tunnel. Therefore, the kinetic energy losses occurring during the flow, which in turn give rise to pressure losses, must be compensated by the pressure increase generated by the fan.

Table 3.2 presents the mechanical and electrical characteristics of the fan used, existing in the Fluid Mechanics and Turbomachinery Laboratory of the Electromechanical Engineering Department (DEM), University of Beira Interior.

Table 3.2: Characteristics of the propulsion elements.

	Features	Fan
Mechanicals	Brand	Aeric
	Model	Golden-Lebey
	Carcass diameter	800 mm
	Hub diameter	300 mm
	Front pass-through area	409000 mm2
	N° of blades	10
Electric	Brand	Somer
	Model	32E
	Rotation speed	1500 rpm
	Operating voltage	380 V
	Power	2,94 kW
	Frequency	50 Hz

3.2 Design and manufacture of the support structure

Some of the constituent elements of our Wind Tunnel were designed and manufactured in a previous project [1], and the remaining elements were designed throughout this work, using Solidworks® 2013 software.

For the support of the main body of the Tunnel, two support structures (Fig. 3.1) were designed and later manufactured, whose characteristics are described in Table 3.3 and can be consulted in Annexes B1 and B2. One of the structures supports the largest diameter section of the diverging element (Fig. 3.1a) and the other supports the stabilisation stator (Fig. 3.1b). The manufacturing process of the structures supporting the main body of the Tunnel consisted of joining several commercial steel tubes by means of a welding process. They were subsequently subjected to an abrasive treatment in order to leave the surface conformed. Finally, they were painted to obtain the final structure. It should be noted that some constraints were foreseen during the transportation of the structure, namely the difficulty of transportation due to the structure's own weight. This and other factors led to the need to manufacture the support structures separately.

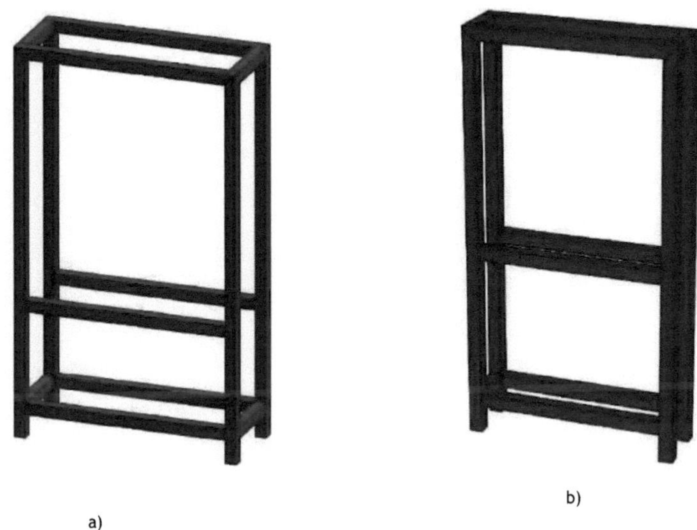

a)

b)

Figure 3.1 - Support structures: a) for divergent nozzle; b) for stabilization stator.

In order to prevent direct contact between the tunnel elements and the support structures of Fig. 3.1, additional MDF wood elements (Fig. 3.2) were designed to provide a connection between the Wind Tunnel and the support structures (see Annexes B3 and B4). The use of the wooden elements will contribute to the reduction of vibrations and degrees of freedom of both the divergent nozzle and the stabilisation stator during the operation of the Wind Tunnel.

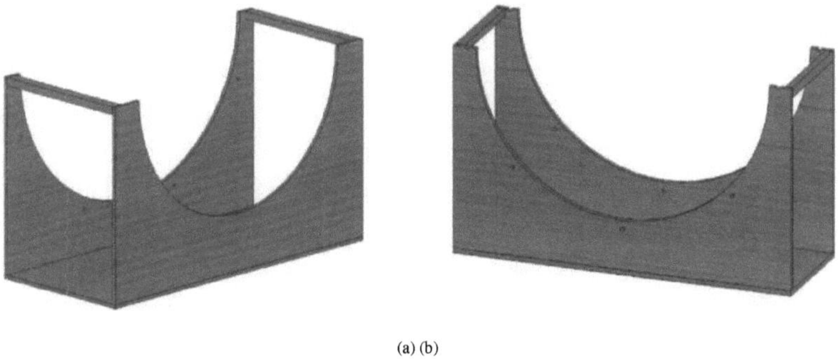

(a) (b)

Figure 3.2 - Wooden elements: a) for divergent nozzle; b) for stabilization stator.

In addition to the support structures to the main body of the Tunnel, a trolley was also used to support the fan. This trolley was designed and manufactured under the previous project of reference [1]. The connection of the fan to the support trolley was made through two rods (Fig. 3.3), manufactured in 4 mm thick commercial steel (see Appendix B5), bolted to the trolley and the fan.

Figure 3.3 - Rod for fan support.

In order to reduce vibrations and to be able to install the instrumentation and measurement equipment without damaging the stainless steel elements of the Tunnel, three MDF-type wooden rings were designed and manufactured, each with a radial hole (Fig. 3.4). The ring with an external diameter of 892 mm and a thickness of 20 mm, positioned between the fan and the constant section duct has an orifice in the upper part with a diameter of 12 mm (see Annex B6). This hole is used for passing the electric motor power supply cable. The ring with an external diameter of 892 mm and a thickness of 10 mm is positioned between the constant section duct and the divergent section duct. This ring has a 4 mm diameter radial hole (see Annex B7), used as a pressure test port. The third ring, with an external diameter of 680 mm and a thickness of 10 mm, is positioned between the divergent section and the stabilising stator (see Annex B8). This ring also has a 4 mm diameter hole in the upper part, used as a pressure tapping.

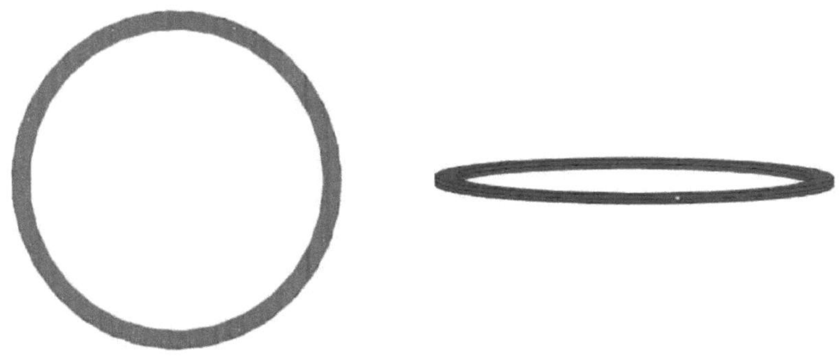

Figure 3.4 - Wooden ring.

For time reasons, it was not possible to manufacture the timber elements shown in Figure 3.2 during the preparation of this study. As an alternative, wooden bars were used to ensure contact between the tunnel elements and the respective support structures, as will be specified in subchapter 3.3.

Table 3.3 summarizes the main dimensions of the elements presented above.

Table 3.3: Characteristics of the tunnel support elements.

Elements	Length [mm]	Height [mm]	Width [mm]
Fan support trolley	890	725	850
Stabilisation stator support structure	800	1525	165
Divergent nozzle support structure	1000	1695	315
Wooden element for support of the stabilization stator	698	401	240
Wooden element for supporting the diverging nozzle	890	520	405
Support rods	160	305	30

The constituent parts of the Tunnel were connected with fastening elements shown in Table 3.4. These elements present different characteristics, according to the place of use.

Table 3.4: Characteristics of the tunnel fastening elements.

Designation	Reference	Quantity
Hexagon Socket Screw	DIN933 M8*30	16
Hexagon Socket Screw	DIN933 M8*45	16
Nut	DIN 934 M8	32
Plated washers	DIN 125 M8	64

To prevent the longitudinal movement of the timber structures, steel bars were designed, which will later be welded on the support structure to the main body of the Tunnel. A total of 8 steel bars 100 mm long, 30 mm wide and 4 mm thick will be fabricated (Appendix B9), as shown in Figure 3.5:

Figure 3.5 - Steel bar.

Before the assembly of the structure-tunnel assembly, computer simulations were performed in order to analyze the stress distribution in the assembly. This procedure aims to choose the ideal position for the steel and timber structures that support the main body of the tunnel. In this case, the Solidworks® 2013 software was used again to perform the simulations. In this context, the effects of the self-weight of the assembly were studied, through a static load analysis. In these simulations several variables were recorded: the von Mises stress distribution, the deformation and the elongation around the assembly. The study of these variables allowed us to select the longitudinal position of the support structures that best fit the criteria imposed by us. Therefore, it was chosen the position that minimizes the efforts in the set, thus allowing reducing the amount of material used. The aesthetic aspect of the set was also one of the factors considered.

Table 3.5 lists the materials used in each of the elements, as well as their yield stress and modulus of elasticity.

Table 3.5: Materials used.

Component	Material used	yield stress l "m-l	Modulus of elasticity [N/,m-l
Stabilisation stator	Stainless steel	1,724 x108	1,93x1011
Divergent nozzle	Stainless steel	1,724 x108	1,93x1011
Constant nozzle	Stainless steel	1,724 x108	1,93x1011
Fan	Cast Iron	1,517x108 *	6,618x1010
Discharge nozzle	Zinc AC 41	3,3x108 *	8,5x1010
Metal stator support	1023 carbon steel	2,827x108	2,05x1011
Divergent metal support	1023 carbon steel	2,827x108	2,05x1011
Fan support	1023 carbon steel	2,827x108	2,05x1011
Rod that connects to the fan	1023 carbon steel	2,827x108	2,05x1011
Stator wooden support	MDF	8x107	3x109
Divergent wooden support	MDF	8x107	3x109
Wooden ring	MDF	8x107	3x109
Fastening elements	Steel	6,204x108	2,1x1011
Metal bar	1023 carbon steel	2,827x108	2,05x1011

***Breakdown** voltage

Once the materials used in the software were chosen, it was possible to identify the forces acting on the system and the fixed elements. Figure 3.6 represents the assembly of the installation, for the model under study.

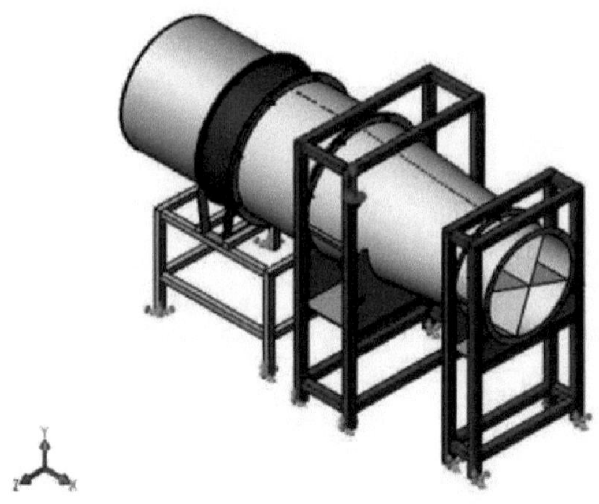

Figure 3.6 - Representation of the simulated model.

In the following step it was defined the type of mesh to be used by the model presented above (Fig. 3.6). Therefore, a triangular mesh type was used (Fig. 3.7). The generated meshes present **479,588** nodes and **250,464** elements. The 4-point Jacobian technique was used, with high quality meshes. During the mesh generation no distorted elements were registered and the time for mesh generation was 1 minute and 31 seconds.

Figure 3.7 - Representation of the meshes in some regions of the model.

Given the geometric differences between the elements of the installation, in relation to their shape and size, the need arose to mesh them independently. Thus avoiding constraints during the meshing process, especially in the fastening elements (bolts, nuts and washers) and in the bolt entry holes. Once the meshes were generated, the simulation was then processed for the values of von Mises stress, elongation, and deformation of the system. Figure 3.8 shows the von Mises stress distribution in the model under study.

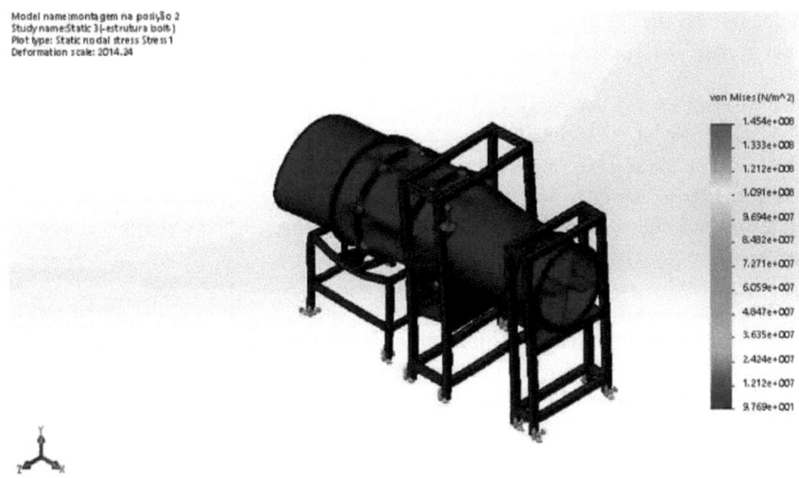

Figure 3.8 - Representation of von Mises stresses in the model.

Figure 3.9 shows the values of the elongation in the computational model under study.

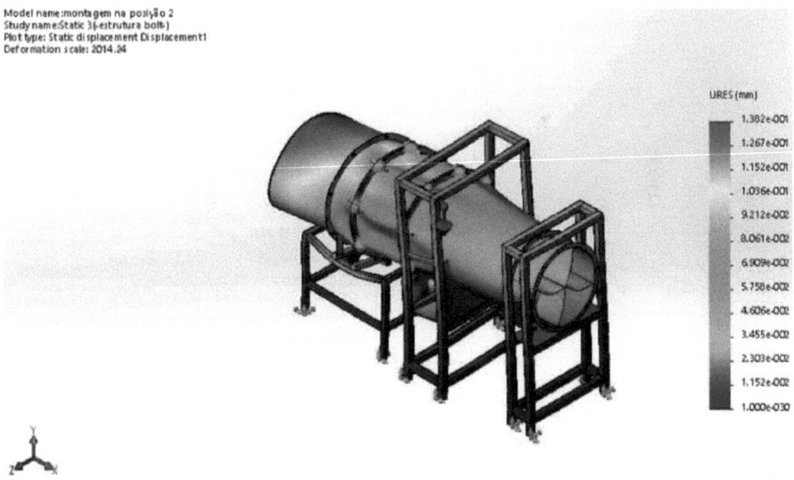

Figure 3.9 - Presentation of the set elongation.

In Figure 3.8 it is notable that the value of the stress in the dark blue zone of the model varies between 97.69 Pa and 12.12 M Pa. These values are low when compared to the values of the rupture stress of the materials used in the project (see Table 3.5). We can then admit that there will be no rupture of the materials used. On the other hand, near the bolted connections (light blue region) the stresses arrive to be higher compared to the remaining parts of the model. However, these stresses are still lower than the yield stress of the materials used.

Through Figure 3.9 we can observe the values of the elongation, which is the result of the action of the self-weight of the elements of the installation. Since most of the Tunnel's weight is concentrated on the rear part of the model, there is a greater elongation in this area, of about 0.138 mm. In other words, we are talking about a maximum 0.138 mm addition to the initial length. This maximum value of elongation does not endanger the structure-tunnel assembly, given the size of the elements that constitute it.

As a complement to the analysis of the elongation, the deformation of the model was also studied. This deformation represents the ratio between the elongation of the material by its initial length, which in this case is almost negligible. The largest deformation of our model is recorded in the bolted joints, as shown in Figure 3.10. The value of this maximum deformation is about 3.840x10-3. It occurs due to the existence of higher stress concentration in these areas of the model under study.

Figure 3.10 - General representation of the model deformation.

In general, through the analysis of the previous parameters, we can state that the assembly presented here ensures safety from the point of view of load distribution along the assembly, as can be seen in the values of von Mises stress (Fig. 3.8). On the other hand, the length of the trolley supporting the fan (Fig. 3.6) makes it impossible to move the support structure to the divergent more to the left. Furthermore, moving this same structure more to the right would imply increasing the height of the wooden support, in order to guarantee the concentricity of the divergent and the other elements. Therefore, increasing the heights of the wooden support would make it more fragile and costly. All the factors presented above contributed to the choice of the geometry presented in Figure 3.6 as being the most appropriate configuration for our installation.

3.3 Installation assembly procedures

As already stated in subchapter 3.2, some elements presented were only designed but not manufactured. These are the cases of the wooden elements and the steel sidebars. However, the assembly of the installation involved a number of steps. These steps are described in the following order:

- Star connection of the three phases and the corresponding neutral to the three-phase electric motor which is coupled to the fan (Fig. 3.11).

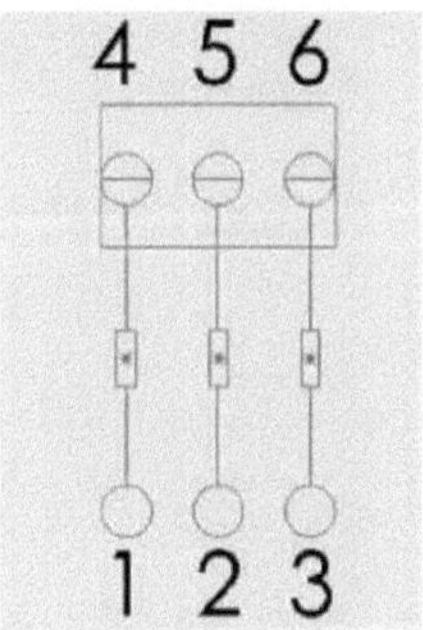

Figure 3.11 - Star Closing of the motor.

- Evaluation of the electrical variables inherent to motor operation, such as: reactive power, active power, apparent power, electric current, voltage, frequency and phase angle. This action aims to avoid possible instabilities in the electrical network and in the motor itself.

- Assembly of the main body of the Tunnel, with the respective wooden rings between the Tunnel elements.

- Use of wooden bars as support elements for the entrance section of the Tunnel (Fig. 3.12)

- Connection of the three pressure inlets to the pressure gauges, located in the inlet sections, in the middle part of the installation and finally in the discharge nozzle. This procedure aims to determine the static pressure in these areas of the Tunnel.

- Assembly of a mobile support mechanism for the Prandtl probe, for reading the dynamic pressure at the entrance and exit of the Tunnel.

Figure 3.12 - Installation side view.

The legend for the figures shown in Figure 3.12 is described below:

1. Stabilisation stator (input);

2. Divergent nozzle (diffuser);

3. Constant section nozzle;

4. Fan;

5. Discharge nozzle.

Chapter 4

Test and calibration of wind tunnel

This chapter presents the tests and calibration of the Wind Tunnel. The equipment used and the respective experimental procedures are also described. Finally, the essential characteristics of the installation are presented, based on the results of the tests performed.

4.1 Measuring equipment

Measurement equipment was used to analyse the flow at the entrance, inside and exit of the Tunnel. The use of this equipment made it possible to obtain information about the distribution of flow velocities at the entrance and exit sections of the installation, as well as to obtain static pressure values inside the Tunnel. Therefore, the following equipment was used in this experiment:

■ Prandtl probe

The Prandtl probe consists of the union of a pitot tube and a static tube, where the pitot tube is mounted co-axially inside the static tube. This equipment is used to measure the total pressure (pitot tube) and static pressure (static tube), of a flow. It can also be used to measure the velocity profile in the boundary layer, as well as the dynamic pressure in the same layer. The tube used is made of stainless steel, with 4 holes of 0.5 mm diameter for measuring the static pressure. This orifice is located 32 mm away from the stagnation orifice. The total length of the probe is 945 mm.

Figure 4.1 shows the Prandtl probe used during the tests. Where the red pressure jack is used to measure the stagnation pressure and the blue pressure jack is used to measure the static pressure. This probe is placed on a wooden board, containing a ruler, which will allow identifying the coordinate of the point to be measured.

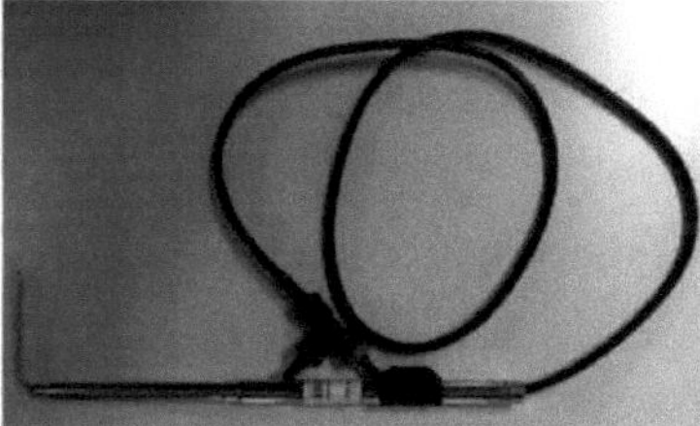

Figure 4.1 - Prandtl probe used for dynamic pressure measurement.

■ Pressure gauge bank

The manometer is one of the oldest devices used for measuring differential pressures, generally easy to construct. This equipment makes it possible to measure the difference in pressure between a reference pressure (generally atmospheric) and a process pressure (generally measured by means of a pressure tapping device). It can also be used to measure dynamic pressures in conjunction with a Prandtl probe. In this case, the two terminals of the manometer are connected to the stagnating and static pressure sockets of the Prandtl probe. Manometers can consist of two glass tubes, parallel and connected at the bottom by another U-shaped tube. The assembly is filled with liquid (usually water) to about half its height. The difference in height of the liquid in the two tubes is measured by means of a scale attached to the tubes (Fig. 4.2).

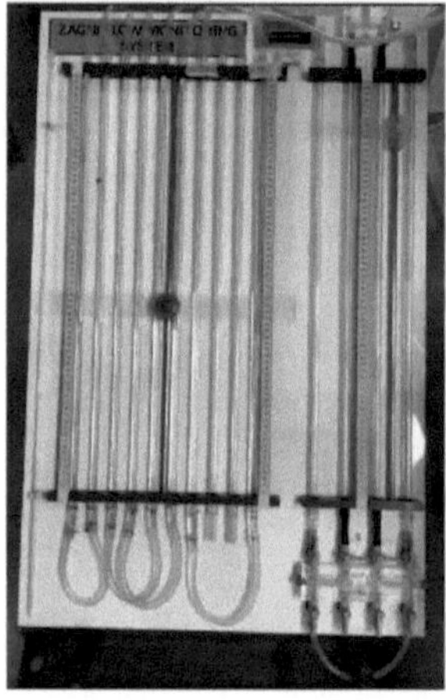

Figure 4.2 - Pressure gauge bank used.

Figure 4.2 shows the manometer bank used during the tests performed. In equation 4.1 Ah represents the difference in height between the two tubes and B represents the inclination of the manometers. The term g represents the acceleration of gravity and p^O represents the density of the liquid (water).p2and p1represent the pressures to be measured.

$$P2\text{-}PI = AhgpH2Osin\ B$$

4.1

A barometer (Fig. 4.3a) and a thermometer (Fig. 4.3b) were used to read atmospheric pressure and temperature, respectively.

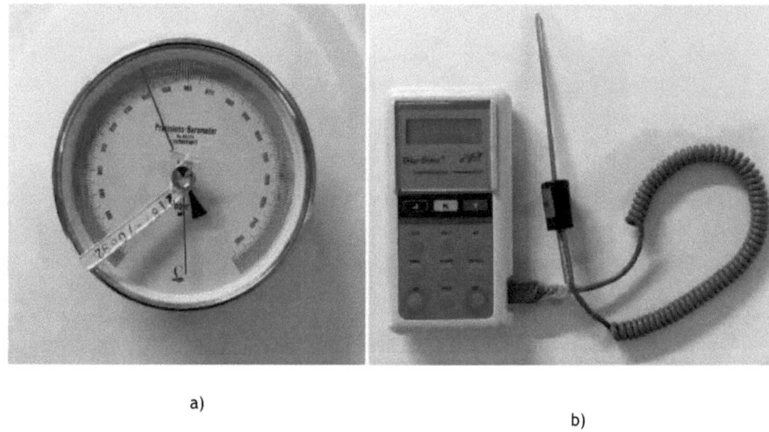

a)

b)

Figure 4.3 - Instruments used for reading the weather conditions: a) Barometer; b) Thermometer.

The register of the motor rotation speed, in rpm, was done through the tachometer shown in figure 4.4.

Figure 4.4 -

4.2 Experimental procedures

Before the tests, all possible steps were taken, with the aim of minimizing systematic measurement errors. To this end, a series of protocols, described below, were followed:

- Verification of the connections between the pressure gauge bank and the three pressure inlets. It should be noted that these connections were made through PVC pipes;

- Opening the laboratory door to avoid recirculation of the air moved by the machine. This measure aims to minimize the influence of recirculation on the stagnation pressure reading;

- Pre- and post-recording of atmospheric pressure and temperature, in order to make a representative average of these conditions throughout the test. This measure aims to minimise the influence of the variation of atmospheric conditions.

Tables 4.1 and 4.2 show the electrical parameters of the motor, measured during tests 1 and 2, respectively.

Table 4.1: Electrical Parameters of the Engine during Test 1.

Recorded Parameters	Recorded value
Operating Voltage U1-4,U2-5,U3-6[V][V	226,1; 226,2; 225,8
Phase current I1,I2,I3[A].	5,42; 5,24; 5,4
Active Power [kW]	2,94
Reactive Power [kVAr]	2,15
Apparent Power [kVA]	3,64
Power Factor: cos ^	0,806

Table 4.2: Electrical parameters of engine during test 2.

Recorded Parameters	Recorded value
Operating Voltage U1-4,U2-5,U3-6[V][V	230; 230; 230
Phase current I1,I2,I3[A].	5,5; 5,24; 5,44
Active Power [kW]	2,95
Reactive Power [kVAr]	2,25
Apparent Power [kVA]	3,71
Power Factor: cos ^	0,796

With the fan in operation, the electrical parameters of the motor were recorded (Table 4.1 and Table 4.2). The first measurements were performed in the entrance section of the tunnel, where, with the aid of a ruler, a Prandtl probe and a manometer, the values of the dynamic height of the water column were measured. The ruler made it possible to identify the distance from the centre to the reading point, where the Prandtl probe was positioned. Figures 4.5 and 4.6 show the breakdown into n points, of the entrance and exit areas of the Tunnel, where n starts at 1. For the first measurement, performed in the centre of the section (where n = 1), the ruler marks 30 cm in the entrance section and 40 cm in the exit section. These values correspond to the radius of each of these sections.

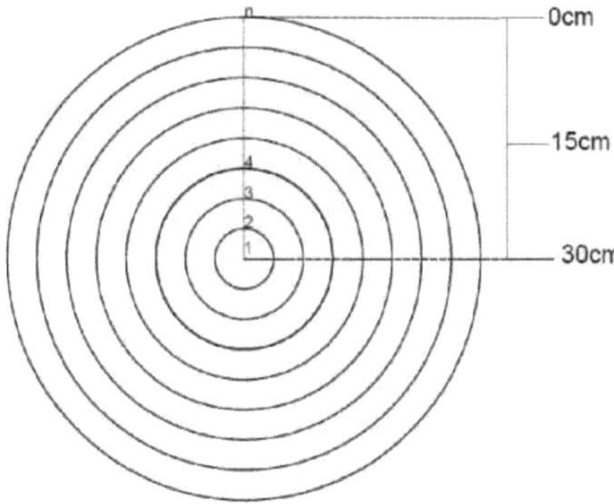

Figure 4.5 - Representation of the points measured in the entrance section of the Tunnel.

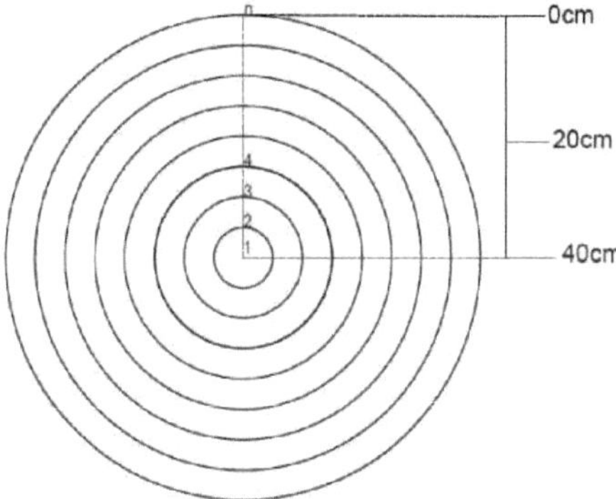

Figure 4.6 - Representation of the points measured in the exit section of the Tunnel.

After the readings at the inlet section (section A), the static pressure values were measured at the three pressure inlets in the Tunnel (Fig. 4.7). Outlet B is located in the dividing region between Tunnel Element 1 and Tunnel Element 2. Inlet C is located in the area between element 2 and element 3. Inlet D is located at the top of the discharge spout (element 5). The last measurements were taken at the exit of the Tunnel (section F) in the same way as at the entrance of the installation.

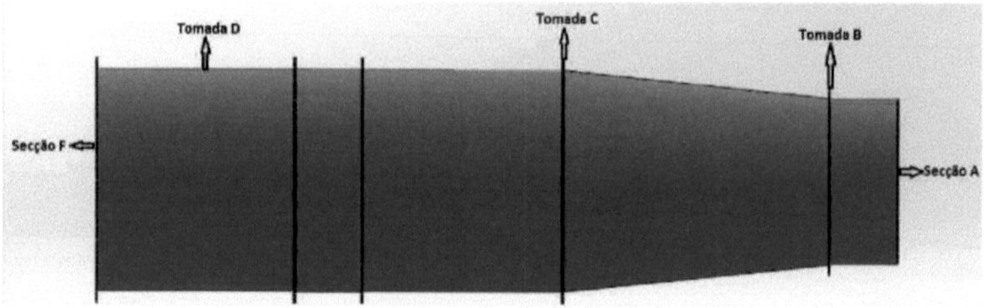

Figure 4.7 - Representation of the tunnel plugs.

4.3 Parameters to be determined

The parameters to be determined aim to identify the main characteristics of the installation. Therefore, the use of the Prandtl probe in the inlet and outlet sections will allow the reading of the static (pj and stagnation (p0) pressures. The static pressure, also called thermodynamic pressure, is the pressure to which the moving fluid particle is subjected. The stagnation pressure, on the other hand, is the pressure that arises when the flow is decelerated to zero velocity by means of a frictionless process. The difference between the stagnation pressure and the static pressure is called the dynamic pressure. From these definitions it is possible to write Bernoulli's equation for the Prandtl probe (equation 4.2).

$$P_0 = P_\infty + \frac{1}{2}\rho_{ar} v^2 \qquad\qquad 4.2$$

In equation 4.2 the term $2 pv2$ then represents the dynamic pressure of the flow, translated by the difference between the total pressure (stagnation) and the static pressure, whose representation is Ap. The variable $_{par}$ represents the density mass of air, calculated from the ideal gases equation (equation 4.3).

$$\rho_{ar} = \frac{P_{atm}}{RT} \qquad\qquad 4.3$$

In equation 4.3 the term $_{patm}$ is the absolute value of the pressure measured by the barometer, T is the temperature obtained by the thermometer in Kelvin and R is the universal constant of ideal gases. From equation 4.2 it is possible to obtain the flow velocity at the measured point (equation 4.4).

$$v = \sqrt{\frac{2\Delta p}{\rho_{ar}}} \qquad\qquad 4.4$$

Given the dynamic pressure measurement methodology, it was necessary to calculate the volume flow rate (Qt) by summing up several elementary flow rates (Qi). Each of these flows is defined as the volume flowing through an elementary control surface per unit time. Since our calculations result in flow velocities, the elementary flow rates are calculated as the product of the average velocity between two points n and n+1 and an elementary control surface area. The elementary control surfaces are the areas enclosed by two consecutive circles (n and n+1) of figures 4.4 and 4.5. In equation 4.5 and 4.6, the expressions for the calculation of the volume flow rate and mass flow rate, respectively, are presented.

$$Q_{(i,i+1)saída} = \sum_{i=1}^{n} \pi v_{mi,i+1}\left(r^2_{i+1} - r^2_{i}\right) \tag{4.4}$$

$$\dot{m} = Q_t \rho_{ar} \tag{4.5}$$

Where i = 1, 2, 3,... n.

The values of the elementary flow rates are determined at the entrance and exit of the Tunnel, where the total volume flow rate (Qt) is the sum of the n elementary volume flows. After the total volume flow rate has been determined, the total mass flow rate (m) is determined at the entrance and exit sections of the Tunnel. It is important to note that m at the entrance and exit of the tunnel should be very close, or even equal, taking into account the law of conservation of mass.

In equation 4.5, vmj,j+1 represents the average velocity of the flow between two consecutive points, located at the boundary of the control surface. The term r is the radius of the circle n control and n, in turn, represents the total number of points measured. The hydraulic power (Ph) of the installation is determined by the product of the total volume flow rate, with the static pressure difference between the entrance and exit of the Tunnel (Ap "B_D).

$$P_h = Q_t \Delta P_{\infty B\text{-}D} \tag{4.7}$$

It is known that in gaseous fluid flows with Mach < 0.3, the flow can be considered incompressible. Therefore, the Mach number (equation 4.8) is an important data to be determined, as a characteristic of the installation.

$$Mach = \frac{v}{\sqrt{KRT}} \tag{4.8}$$

In equation 4.8 the term K represents the adiabatic constant of air.

Another important parameter to be determined is the Reynolds number, because it allows identifying if the flow regime is laminar or turbulent. The Reynolds number was determined for the entrance and exit sections of our Tunnel through equation 4.9.

$$R_e = \frac{vD}{\upsilon}$$

4.9

In this equation v represents the kinematic viscosity of the fluid (air) and D represents the diameter of the section to be considered.

4.4 Essential characteristics of the installation

In order to determine the essential characteristics of the installation, two tests were carried out. In the first test, measurements of the dynamic water column heights were taken at 7 points in the inlet section and at 6 points in the outlet section. In the second test 11 measurements were taken at the inlet and outlet of the facility. During both tests, the static pressure distribution along the Tunnel was also measured, through pressure inlets B, C and D. Therefore, through the results obtained from these tests it was possible to determine the essential characteristics of the installation.

For the first test, the values for atmospheric pressure and temperature are $_{patm} = 943$ mbar $\pm 0,25$ mbar and T = 19,5 $°C \pm 0,1$ T. The pressure gauge used was inclined at an angle of 33° to the horizontal plane in order to obtain a better accuracy during readings. The rotational speed of the electric motor, as measured by tachometer, is 1451 rpm. The air density was determined by equation 4.3, using the values of pressure and atmospheric temperature, having been obtained 1.1227 kg/m3.

Considering the sweep of figure 4.5, the dynamic heights of the water column in 7 points were measured with the Prandtl probe. The values obtained are presented in table 4.3.

Table 4.3: Dynamic heights at the tunnel entrance for test 1.

Points	r [cm]	Ahinput
1	30	27,5
2	25	31
3	20	30,5
4	15	31,5
5	10	33
6	5	26
7	0	0

With the values presented in table 4.3, the dynamic pressure, the flow velocity, the elementary volume flows and the mass flow rate at the tunnel entrance were calculated. These values are described in table 4.4.

Table 4.4: Velocity and flow distribution at the tunnel entrance for test 1.

Ap [Pa]	V [m/s]	Vmi [m/s]	Vm [m/s]	Qi [$m^{3/s}$]	Q [$m^{3/s}$]	m [kg/s]
146,93	16,1782	16,6775		1,4408		
165,6302	17,1768	17,1073		1,2092		
162,959	17,0377	17,1763		0,9443		
168,3012	17,3148	17,5186	16,860	0,6879	4,73823	5,31983
176,3159	17,7223	16,7265		0,3941		
138,9156	15,7307	7,8654		0,06177		
0	0	0		0		

For the outlet section the measurements were made in 6 points, using the Prandtl probe, according to the sweep of figure 4.6. Table 4.5 presents the values of dynamic heights of water column for the respective measured points.

Table 4.5: Dynamic heights at the tunnel exit for test 1.

Points	r [cm]	Ahinput
1	40	1
2	32	5.5
3	24	9
4	16	31,5
5	8	27
6	0	0

With the values obtained in Table 4.5, important parameters were calculated for characterising the flow at the Tunnel outlet. These parameters are: flow velocity, elementary volume flow rates and mass flow rate. These values are presented in table 4.6.

Table 4.6: Velocity and flow distribution at the tunnel exit for test 1.

Ap [Pa]	V [m/s]	Vmi [m/s]	Vm [m/s]	Q [m3/s]	Qt [m3/s]	m [kg/s]
5,3429	3,0851	5,1601		0,9337		
29,3859	7,2551	8,2451		1,1604		
48,0862	9,2552	13,2850	10,584	1,3355	4,59657	5,16078
168,3018	17,3148	16,6726		1,0057		
144,2585	16,0304	8,0152		0,1612		
0	0	0		0		

The pressure taps B, C and D (Fig. 4.7) allow obtaining the values of the relative static pressure in different zones of the installation. Table 4.7 shows these values.

Table 4.7: Relative static pressures for test 1.

Socket B		Socket C		Socket D	
AhB	p.B [Pa]	AhC	p C. [Pa]	AhD	p.D [Pa]
-107,95	-576,767	-55	-293,8599	0	0

From equations 4.8 and 4.9, the Mach and Reynolds numbers in the inlet section and in the interior part of element 3 were determined. For the calculation of the Mach and Reynolds numbers in the inlet section, the average velocity **vm** presented in table 4.4 was used. While for the interior of element 3, the velocity was determined by dividing the flow of the installation by the respective area, whose diameter is 0.8 m. The values of hydraulic power and hydraulic efficiency of the installation are also presented. For this purpose it was assumed that the total volumetric flow rate at the entrance (table 4.4) is equivalent to the flow rate of the installation.

The hydraulic power was obtained through equation 4.7, considering Ap^B-D as being the difference between the static pressures measured at outlets B and D and Q the total volume flow rate of the installation. Through the ratio between the hydraulic power by the electric active power of the motor (table 4.1), the hydraulic efficiency of the installation was obtained. Table 4.8 shows the flow characteristics (Mach and Reynolds), as well as the hydraulic power and the hydraulic efficiency of the installation.

Table 4.8: Main characteristics of installation for test 1

Axe	Mach3	ReA	Re3	Ph [kW]	nh [%]
0,049	0,027	6,74x105	5,03x105	2,733	93

In Table 4.8, indices A and 3 represent values calculated at the inlet sections and inside the constant pipe, respectively (Fig. 4.7).

In the second test, with the manometer in the same position, the pressure and the atmospheric temperature measured on the spot were: $_{patm}$ = 945 mbar ± 0,25 mbar and T = 19,9°C ± 0,1 °C, respectively. The rotational speed of the electric motor, measured by tachometer, is 1456 rpm. The air density under these conditions is 1,123 kg/m3. In this test, the inlet section was divided into 11 points, according to the sweep of figure 4.5. Subsequently, measurements of the dynamic water column heights were made and the results of the measurements are presented in table 4.9.

Table 4.9: Dynamic heights at the tunnel entrance for test 2.

Points	r [cm]	Ahinput
1	30	28,5
2	27	28,5
3	24	28,5
4	21	29
5	18	29
6	15	27
7	12	27,5
8	9	28
9	6	28,5
10	3	35,5
11	0	0

Based on these measurements, important variables for the characterization of the Tunnel entrance section were determined, whose results are shown in Table 4.10.

Table 4.10: Velocity and flow distribution at the tunnel entrance for test 2.

Ap [Pa]	V [m/s]	Vmi [m/s]	Vm [m/s]	Q [m3/s]	Q [m3/s]	m [kg/s]
152,273	16,465	16,465		0,88452		
152,273	16,465	16,465	15,08840	0,79141	4,63198	5,20355
152,273	16,465	16,537		0,70135		
Ap [Pa]	**Vj [m/s]**	**Vmi [m/s]**	**Vm [m/s]**	**Qi [m3/s]**	**Qt [m3/s]**	**m [kg/s]**
154,944	16,609	16,609		0,61048		
154,944	16,609	16,317		0,50749		
144,258	16,026	16,099		0,40969		
146,929	16,173	16,267	15,08840	0,32155	4,63198	5,20355
149,601	16,319	16,392		0,23117		
152,273	16,465	17,420		0,14777		
189,673	18,376	9,188		0,02598		
0	0	0		0		

In the outlet section of the Tunnel, the values of the dynamic water column heights, measured through the Prandtl probe, are presented in Table 4.11.

Table 4.11: Dynamic heights at the tunnel exit for test 2.

Points	r [cm]	Ahinput
1	40	4
2	36	4
3	32	6,5
4	28	6,5
5	24	6
6	20	28,5
7	16	28
8	12	27,5
9	8	21
10	4	13,5
11	0	0

The variables for characterization of this section, namely velocity distribution, average velocity, elementary volume flow rates and mass flow rate, are presented in table 4.12.

Table 4.12: Velocity and flow distribution at the tunnel exit for test 2.

Ap [Pa]	Vj [m/s]	Vmi [m/s]	Vm [m/s]	Qi [m³/s]	Qt [m³/s]	m [kg/s]
21,372	6,168	6,168		0,58910		
21,372	6,168	7,016		0,59950		
34,729	7,863	7,863	10,004	0,59286	4,86381	5,46398
34,729	7,863	7,709		0,50373		
32,057	7,555	12,009		0,66405		
152,273	16,465	16,392		0,74157		
Ap [Pa]	**Vj [m/s]**	**Vmi [m/s]**	**Vm [m/s]**	**Qi [m³/s]**	**Qt [m³/s]**	**m [kg/s]**
149,601	16,319	16,247		0,57165		
146,929	16,173	15,153		0,38085		
112,201	14,133	12,733	10,004	0,19200	4,86381	5,46398
72,129	11,332	5,666		0,02848		
0	0	0		0		

The values of the static pressures measured through the three inlets, during the first test (table 4.7), remained constant in the second test. However, due to the change in the value of the total volume flow in the inlet section, the hydraulic power and the hydraulic efficiency obtained in this test (table 4.13) differ from the values obtained in the first test. The procedure for calculating the Reynolds and Mach numbers are the same as those used in the first test.

Table 4.13: Main characteristics of the installation for test 2.

Axe	Mach3	ReA	Re3	Ph [kW]	Oh [%]
0,044	0,0268	6,03x105	4,91x105	2,67	90,6

In the outlet section, the obstruction of the flow by the rotor block and the adherence conditions of the boundary layer of the particles in contact with the rotor gave rise to the appearance of secondary flow in this section of the Tunnel. The secondary flow is characterized by the presence of vortices downstream of the rotor block and consequently the deceleration of the flow. Thus, points 1, 2 and 3 (Fig. 4.8a) and 1, 2, 3 and 4 (Fig. 4.8b), close to the rotor block ($_{dbl\ rotor\ block}$ = 30 cm), were affected by the presence of vortices.

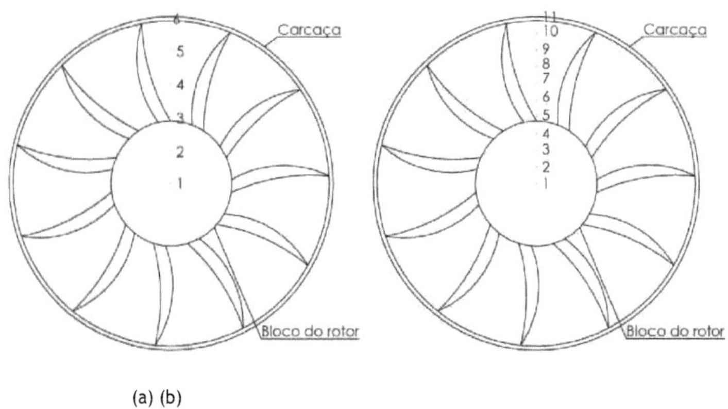

(a) (b)

Figure 4.8 - Points measured at the Tunnel outlet section (downstream of the rotor): a) Test 1; b) Test 2.

In Figure 4.8a the flow velocity increases from point 1 to point 4, where it reaches the maximum value of 17.31 m/s. As we approach the Tunnel walls, i.e., from point 4 to point 6 the velocity decreases to zero (Table 4.6). This occurs due to the adherence conditions of the boundary layer. For the same reasons, although with some fluctuations, the flow velocity in test 2 (table 4.12) increases from point 1 to point 6 (Fig. 4.8b), where it reaches the maximum value of 16.465 m/s. Then, near the Tunnel walls (point 11 of Fig. 4.8b) the velocity decreases to zero.

In the entrance section, the presence of the cross of element 1 causes a decrease of the flow velocity in the centre of the Tunnel. Instead the maximum velocity occurs at point 5 of Table 4.4, decreasing again near the wall. This happens due to the boundary layer adherence conditions at the cross of Tunnel element 1. Figures 4.9 and 4.10 show the velocity distribution at the entrance and exit of the facility for test 1, where the vertical axis represents the position of the Prandtl probe and the horizontal axis represents the velocities of the flow.

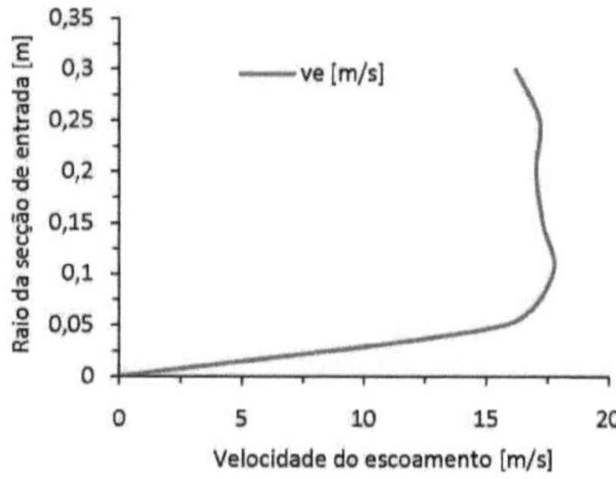

Figure 4.9 - Distribution of flow velocities at the entrance of the Tunnel for test 1.

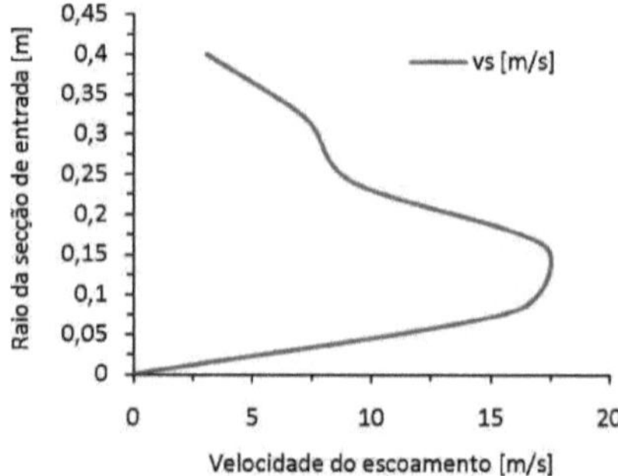

Figure 4.10 - Distribution of flow velocities at the exit of the Tunnel for test 1.

The velocity distribution curve of the flow presented in Fig. 4.9 is typical of turbulent flow. For the curve at the outlet (Fig. 4.10) it is notable the influence of the appearance of secondary flow. This behavior is represented by the distortion of the velocity profile curve at the outlet.

As can be seen in Tables 4.9 and 4.13, the calculated Reynolds values are higher than 2300, which means that the flow is turbulent in both sections of the Tunnel. As a guarantee that the flow can be treated as incompressible, the Mach numbers determined in both tests are lower than 0.3.

The difference between the mass flow rates at the entrance and exit of the Tunnel is 2.89% in Test 1 and 4.77% in Test 2. Several factors underlie this difference, among them is the appearance of secondary flow in the exit section of the Tunnel and systematic errors during the measurement.

When comparing the hydraulic power of the installation with the electrical active power of the fan, a deviation of 7% is found for test 1 and 9% for test 2. These differences include internal losses, e.g. during fluid contact with the rotor blades, etc. They also include mechanical losses in the motor bearings.

The difference between the inlet mass flow rates obtained in tests 1 and 2 is 2.19%, while for the outlet mass flow rates this difference is 5.55%. Factors such as the difference in atmospheric conditions and measurement errors are at the base of this disparity. Therefore, the use of computer simulations becomes indispensable for the final characterisation of the installation.

Chapter 5

Computational simulation of the flow in the tunnel

This chapter describes the steps of the computational simulation of the flow along the Wind Tunnel. Therefore, the geometry of the Wind Tunnel computational domain was initially designed in 3D, using Solidworks® 2013 software. Then the geometry was imported into Ansys-Fluent® 2016 software for the flow analysis along the tunnel.

The software used in this analysis allows solving the problem under study according to the sequence shown below:

- Problem Identification:

  1- Set targets;

  2- Identification of the computational domain;

- Pre-processing:

  3- Geometry;

  4- Knit;

  5- Physical properties;

  6- Simulation settings;

- Simulation:

  7- Computational simulation;

- Post-processing:

  8- Analysis of results;

  9- Update the model (return to preprocessing).

Updating the model consists of returning to the preprocessing step, where the user has the ability to edit the model's geometric characteristics, mesh properties, physical properties of the model, and simulation settings.

5.1 Definition and meshing of the computational domain

For our model a 3D axisymmetric computational domain was used. In order to simplify the problem solving, only a fraction of the domain was solved (Fig. 5.1). The simplified computational domain that was chosen represents a longitudinal section of the wind tunnel. This choice is justified given the symmetry of the flow characteristics in circular-section tunnels, relative to a plane that crosses the longitudinal axis. The domain was divided into four regions, namely, the inlet region, the outlet region, the symmetry region and the tunnel walls.

Subsequently, the boundary conditions were imposed on these regions. It should be noted that among the constituent elements of the tunnel, the discharge nozzle is not part of the computational model presented for the simulations.

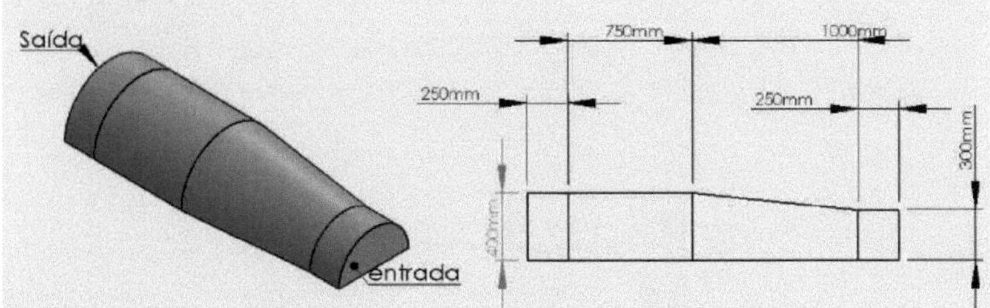

Figure 5.1 - Computational domain.

The type of mesh to be used is closely dependent on the geometry of the computational domain. Therefore, for simple and regular geometries, it is recommended to use structured meshes. In the case of a more complex geometry, structured meshes can cause problems in the physical solution of the analysis that is intended to be performed. As an alternative, in these cases it is recommended to use unstructured meshes. According to our computational domain were then tested two different types of meshes: structured meshes (quadrilaterals) and unstructured meshes (tetrahedral). Between the two types only one was used due to the quality of the generated mesh. To characterize the quality of the generated mesh, the program considers some important variables during the meshing process: the distortion angle and the aspect ratio.

Ansys-Fluent® 2016 uses two expressions to define the distortion angle, whose equations are represented below:

$$\dot{m}=Q_t \rho_{ar}$$

5.1

$$Skn=\max\left[\frac{\theta_{máx}-\theta_{min}}{180-\theta_e}, \frac{\theta_e-\theta_{min}}{\theta_e}\right]$$

5.2

Expression 5.1 is used for triangular and tetrahedral meshes, while expression 5.2 is used for quadrilateral and hexahedral meshes. The variable Toc represents the optimal cell size and Tc represents the cell size. For expression 5.2, 0e represents the slope of the cell. The program recommends that the angle of distortion be less than 0.9 to avoid poor meshing quality, which in turn can lead to incorrect and slowly converging physics solutions.

Figure 5.2 presents the values of the angle of distortion recommended by the program used. Classified from excellent to unacceptable.

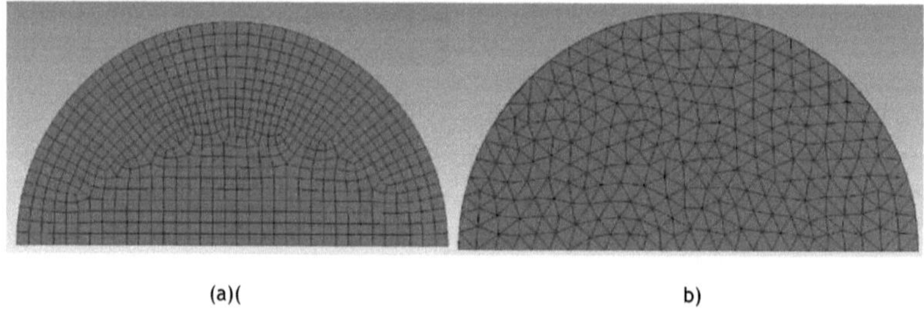

Figure 5.2 - Skn values recommended by the program.
Source: Ref. A10

The aspect ratio is defined as the quotient of the larger side of the mesh, by the smaller side. It is recommended that its value be less than 40, for flows aligned with the model geometry.

Figure 5.3 shows the two types of meshes generated in the model output section.

Figure 5.3 - Meshes generated at the model output: a) quadrilateral meshes; b) Tetrahedral meshes.

Table 5.1 gathers the characteristics of the quadrilateral and tetrahedral meshes tested in our computational model. Based on this table, it is possible to determine the quality level of the generated meshes.

Table 5.1: Details of the meshes generated.

Mesh type	N° of elements	No. of nodes	Size [m]		*Skewness*		Aspect ratio	
			Min	Max	Min	Max	Min	Max
Quadrilatera	79729	86532	1-10-3	2-10-2	1,306-10-	0,435	1,0033	4,4477
Tetrahedral	118124	22311	1-10-3	6,4-10-2	1,233-10-3	0,849	1,1732	8,7767

According to the *Skewness* values presented in table 5.1 and the values recommended by the program, indicated in figure 5.2, the quadrilateral meshes present better quality in comparison with the tetrahedral ones. This analysis shows that the values of Skn for the quadrilateral meshes are in a range called very good, ensuring a lower distortion of the elements. For this reason and the fact that the flow is aligned with the geometry, we choose the quadrilateral meshes. In Fig. 5.4 shows a general view of the computational domain, after the mesh generation.

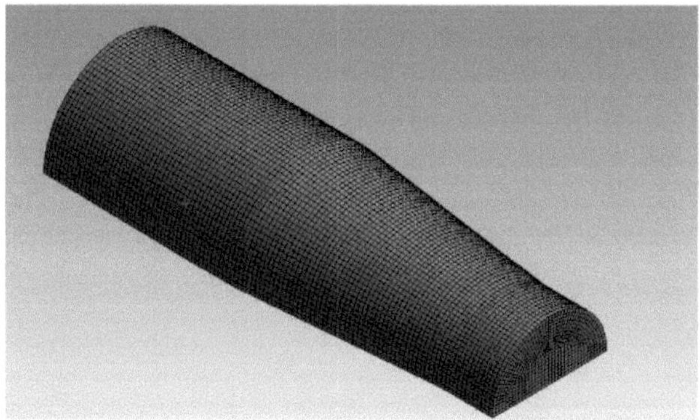

Figure 5.4 - Overview of the quadrilateral mesh chosen.

In summary, the type of mesh to be used for a given problem can dictate the success or failure in obtaining the computational solution. However, the mesh must be fine enough to ensure adequate solution to the problem.

5.2 Governing equations and turbulence model

The mathematical model to be used is based on the governing equations of fluid dynamics for a specific problem. In this case, the tests performed show that there are no large variations in density, implying an incompressible flow. Therefore, simulations for a non-viscous incompressible flow and simulations for a viscous incompressible flow were performed.

The model for non-viscous flow was performed using the Euler equations in combination with the continuity equation:

$$\frac{\partial \rho}{\partial t} + \nabla \cdot \left(\rho \vec{V} \right) = 0 \qquad\qquad 5.3$$

$$\rho \frac{d\vec{V}}{dt} = \rho \vec{g} - \nabla p \qquad\qquad 5.4$$

Where V=(u, v, w)

Equations 5.3 and 5.4 represent the continuity and Euler equations, respectively. Where ρ is the density mass of the fluid, t represents time, $\vec{V}$ is the velocity vector of the flow, g represents gravity acceleration, p is the pressure and the components u, v and w represent the velocity in the x, y and z axes, respectively. Since the flow is incompressible ($\rho = cte$), the continuity equation is reduced in the following form: $\nabla\vec{V}=0$.

For incompressible viscous flow, the continuity equation was combined with the system of nonlinear second order differential equations, called Navier-Stokes equations (equation 5.5):

$$\rho\frac{d\vec{V}}{dt} = \rho\vec{g} - \nabla p + \mu\nabla^2\vec{V} \tag{5.5}$$

The solution of the above equations will allow obtaining the velocity and pressure fields of the partial differential nonlinear system. For this case, the equations were solved considering flow in permanent regime, i.e., in equations 5.4 and 5.5 the term rIV- =0. On the other hand, for turbulence effects the standard k-c turbulence model was considered.

The standard k-c model was proposed in 1974 by B. E. Launder and D. B. Spalding [A11]. It is a semi-empirical model based on two model equations, namely the equation for the kinetic energy of turbulence k and the equation for its dissipation rate c. In this model, the transport equation for turbulence kinetic energy is derived from the exact equation, while the equation for the dissipation rate was obtained using physical reasoning. For the application of this model it is assumed that the flow is completely turbulent and molecular effects are negligible.

$$\begin{cases} \dfrac{\partial(\rho k)}{\partial t}+\dfrac{\partial(\rho k u_i)}{\partial x_i}=\dfrac{\partial}{\partial x_j}\left[\left(\mu+\dfrac{\mu_t}{\sigma_k}\right)\dfrac{\partial k}{\partial x_j}\right]+G_k+G_b-\rho\epsilon-Y_M+S_k \\[3ex] \dfrac{\partial(\rho\epsilon)}{\partial t}+\dfrac{\partial(\rho\epsilon u_i)}{\partial x_i}=\dfrac{\partial}{\partial x_j}\left[\left(\mu+\dfrac{\mu_t}{\sigma_c}\right)\dfrac{\partial\epsilon}{\partial x_j}\right]+C_{1c}\dfrac{\epsilon}{k}(G_k+C_{3c}G_b)-C_{2c}\rho\dfrac{\epsilon^2}{k}+S_c \end{cases} \tag{5.6}$$

The variable Gk represents the turbulence kinetic energy generation due to the mean velocity gradient. The variable Gb is the turbulence kinetic energy generation caused by buoyancy. The variables S_e and S_k represent the source terms. The variable YM represents the buoyant dilatation in compressible turbulence, which is a function of the overall dissipation rate. In the system of equations 5.6, the turbulent viscosity is $_{pt}= pCM(k2/c)$.

The model has the following values for its constants: $_{ok=1}$, $_{otó,1=2}$, $_{oe=1}$,3, $_{C1e=1}$,44, C2e=1,92 and C^=0,09.

5.3 Imposition of boundary conditions and method of resolution

There are two types of boundary conditions, namely Dirichlet conditions and Neuman conditions. Dirichlet conditions can be simply defined as the transport of properties ^ around the boundary. Neuman conditions involve an analysis of the properties around the boundary, given by |n = 0. Generally, Dirichlet conditions are used to define the inlet flow, and Neuman conditions are used for the outlet flow.

The specifications of the boundary conditions depend on the problem. Therefore, for our model, static pressure values were imposed at the tunnel entrance and exit. Initially these conditions were imposed for a non-viscous flow and later for a viscous flow. For both cases it was considered incompressible flow, in permanent regime. The pressure values used in the boundaries were obtained from the tests of the installation (Chapter 4). For both non-viscous and viscous flow, the pressure at the tunnel entrance is atmospheric ($_{pe=}$ 0 Pa), whereas at the Tunnel exit the pressure is -293.859 Pa (pressure read through outlet C), assuming that the fan discharges to atmosphere. The outlet boundary is limited upstream of the fan, admitting constant head losses in the Tunnel element 3.

The process for obtaining the computational solution consists of two stages. The first stage involves the discretization stage, that is, the conversion of the differential equations

65

partial and auxiliary equations in systems of discrete algebraic equations. The software used performs this step using the Finite Volumes method. In this method, the computational domain is divided into a finite number of control volumes. Expression 5.7 represents the model equation for the Finite Volume method.

$$\frac{\partial}{\partial t}\int_V \rho\varphi\,dV + \oint_A \rho\varphi\,dA = \oint_A \Gamma_\varphi \nabla\phi\cdot dA + \int_V S_\varphi\,dV \qquad\qquad 5.7$$

In equation 5.7 the first term represents the unsteady term along the control volume, the second term represents the convection on the control surface, the third term represents the diffusion term along the control surface and the last one represents the source term. The variable A represents the surface area of the control volume and V the control volume itself. On the other hand, the variables r^ and S^ represent the diffusion coefficient and source term of the property ^, respectively.

The general discretization equation around a one-dimensional control volume is given by formula 5.8:

$$a_p\varphi_p = a_E\varphi_E + a_W\varphi_W + b$$

Onde:

$$a_E = \frac{\Gamma_E A_E}{\Delta V \delta x_E} \qquad\qquad a_W = \frac{\Gamma_W A_W}{\Delta V \delta x_W} \qquad\qquad a_p = a_E + a_W \qquad\qquad b = S_\varphi$$

$$\qquad\qquad 5.8$$

The indices W and E represent, respectively, the boundary nodes west and east of the central node P. The variable δx represents the distance between considered boundary node and the central node P.

The second step to obtain the numerical solution referring to the algebraic equations consists in choosing a direct or iterative numerical method. Direct methods, such as Gaussian elimination and Thomas algorithm, are some of the methods used for solving systems of linear equations. Unfortunately, a large part of computational fluid dynamics (CFD) problems are usually described by systems of nonlinear equations.

The numerical methods of Jacobi and Gauss-Siedel are some of the iterative methods used. The iterative method initially assumes an arbitrary solution and then uses the system equations to systematically improve the solution until the desired convergence is achieved.

Fluent has two methods for solving CFD-type problems, namely the *Pressure-based method* and the *Density-based* method. For our model, we have chosen the pressure-based method. This method is applied to both incompressible flows at low velocity and compressible flows at high velocity. The density based method is generally applied when there is a strong interdependence between the density mass, energy, momentum, etc. This method widely used for supersonic flows.

Figure 5.5 illustrates the steps for obtaining the computational solution based on the Finite Volume and Finite Difference methods.

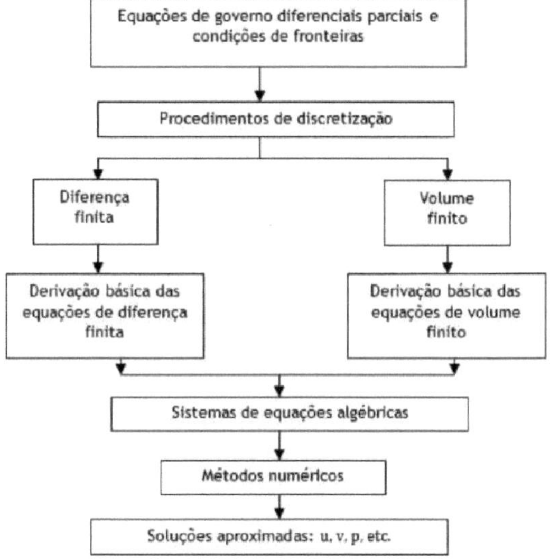

Figure 5.5 - Procedures for computational solution.

Source: Ref.22 (Computational Fluid Dynamics, 2013)

The pressure-based method has two different solving algorithms, *Segregated* and Coupled. The former solves the pressure, pressure correction and momentum equations sequentially (iteratively), while the latter solves the pressure, momentum and pressure correction equations simultaneously (directly). For our study, we chose the Segregated algorithm because it requires less memory compared to the Coupled algorithm. However, it requires more time for convergence to be achieved.

To link the pressure with the velocity a pressure field was used, in order to guarantee the conservation of the continuity equation. For this, it was used the implicit iterative method called SIMPLE *(Semi-Implicit-Method for Pressure-Linked Equations)*. This method consists of initially assuming a pressure field and using it to solve the momentum equation. From the continuity equation the pressure correction equation is deduced, which in turn is used to update the velocity and pressure fields.

The flowchart in Figure 5.6 represents the steps to obtain the solution of the pressure and velocity fields, from the momentum equations:

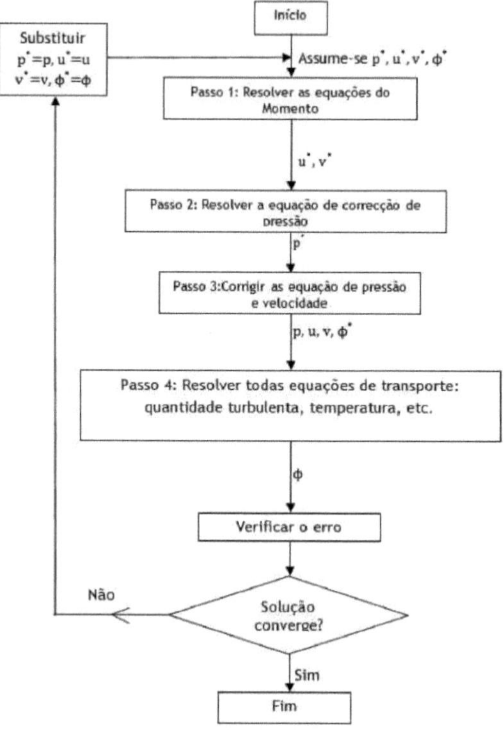

Figure 5.6 - Procedures for the SIMPLE method in the two-dimensional plane.

Source: Ref.22 (Computational Fluid Dynamics, 2013)

The equations of momentum in x and y, are solved using the mathematical expressions shown below:

$$a_p^u u_p^* = \sum a_{nb}^u u_{nb}^* - \frac{\partial p^*}{\partial x} \Delta V + b$$

5.9

$$a_p^u v_p^* = \sum a_{nb}^u v_{nb}^* - \frac{\partial p^*}{\partial y} \Delta V + b' \qquad 5.10$$

In equations 5.9 and 5.10 the AV term represents the change in the control volume, the expressions unb and vnb represent the velocities at the nodes neighbouring the central node P, the term anb represents the neighbourhood coefficient and the term b represents the source term previously presented. The pressure correction p , mentioned in figure 5.6, represents the difference between the actual pressure p of the field and the assumed pressure p*. The same is true for the velocities. The correction velocities are determined by equations 5.11 and 5.12.

$$a_p^u u_p' = \sum a_{nb}^u u_{nb}' - \frac{\partial p'}{\partial x} \Delta V \qquad 5.11$$

$$a_p^u v_p' = \sum a_{nb}^u v_{nb}' - \frac{\partial p'}{\partial y} \Delta V \qquad 5.12$$

During the solving step with the SIMPLE method, the p term is used only as a numerical artifice whose purpose is to accelerate the convergence of the velocity equations. Therefore, equations 5.13 and 5.14 allow the exact values of velocities to be determined, based on the pressure correction values and the assumed values of the velocities.

$$u_p = u_p^* - D^u \frac{\partial p'}{\partial x} \qquad 5.13$$

$$v_p = v_p^* - D^v \frac{\partial p'}{\partial y} \qquad 5.14$$

Where
$$D^v = \frac{\Delta V}{a_p^v} \qquad\qquad D^u = \frac{\Delta V}{a_p^u}$$

The pressure and velocity fields are progressively improved through the iteration process until convergence is achieved. Therefore, for our model we have chosen different levels of convergence. Table 5.2 presents the convergence levels assumed for the different variables.

Table 5.2: Levels of convergence.

Waste	Convergence Level
Continuity	10-5
Speed - x	10-3
Speed - y	10-3
Speed - z	10-3
Energy	10-3
k	10-3
e	10-3

During the iterative process different relaxation factors (table 5.3) were used to stabilize and speed up the iteration process, taking into account the variable in question.

Table 5.3: Relaxation factors.

Flow	p	P	q	k	e	Mt
Non-viscous	0,3	1	0,7	-	-	-
Viscous	0,3	1	0,7	0,8	0,8	1

q - quantity of movement.

5.3 Analysis and validation of results

The results obtained in the computer simulation were used to validate the values obtained in the tests. In this subchapter, the results of the simulations are presented for viscous and non-viscous flow, and then these values are compared with the values obtained in the experimental tests. Figure 5.7 presents the longitudinal values (abscissa axis) of our computational model.

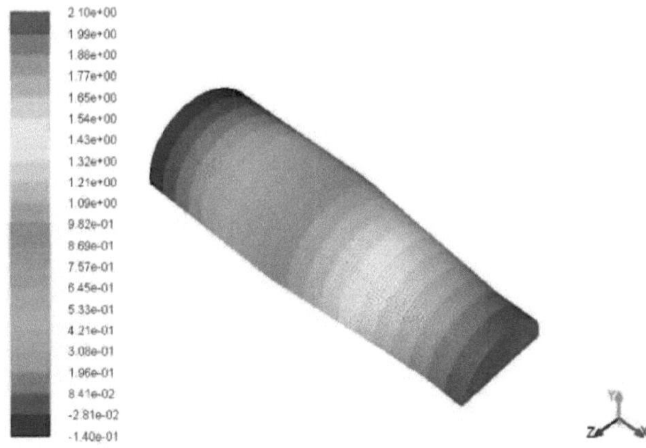

Figure 5.7 - Coordinates of the computational domain with respect to the abscissa axis (X), in [m].

Figure 5.7 shows the coordinates of the computational domain with respect to the abscissa (X) axis. In this figure, the input section is located at X = 2.10 m and the output section is located at X = -0.14 m.

Figure 5.8 - Coordinates of the computational domain with respect to the ordinate axis (Y), in [m].

Figure 5.8 illustrates the coordinates of the computational domain of our model with respect to the ordinate (Y) axis. Considering the longitudinal section, the plane of symmetry is located at Y = 0 (dark blue region).

Therefore, Figure 5.9 shows the distribution of relative static pressures along the Tunnel.

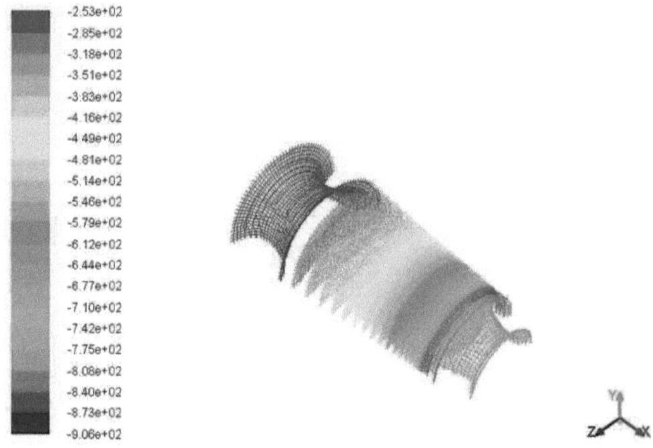

Figure 5.9 - Static pressure distribution in the Tunnel, for non-viscous flow, in [Pa].

In Figure 5.9 it is visible the growth of the absolute value of the static pressure as we move away from the inlet section. This pressure grows due to the increase of the Tunnel section as the flow advances towards the outlet section. Conversely, the dynamic pressure decreases in the direction of flow. The values of the dynamic pressure are shown in Figure 5.10.

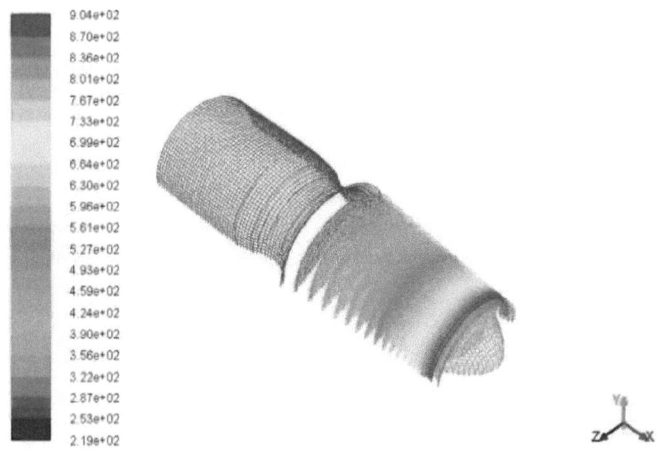

Figure 5.10 - Dynamic pressure distribution in the Tunnel, for non-viscous flow, in [Pa].

Figure 5.11 shows the stagnation pressure values in different zones of our computational model.

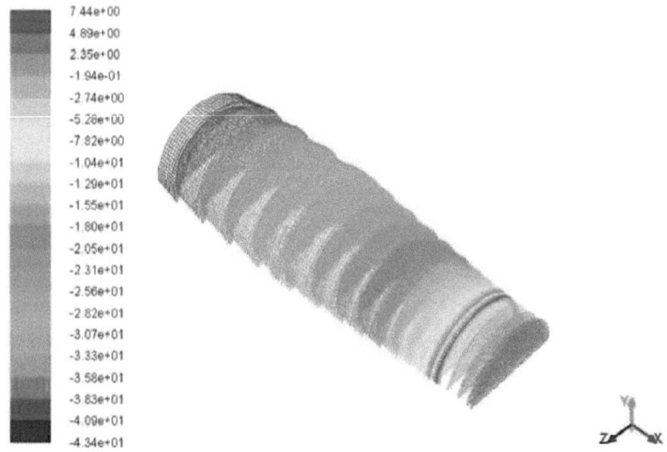

Figure 5.11 - Stagnation pressure distribution in the Tunnel, for non-viscous flow, in [Pa].

Figure 5.10 shows the dynamic pressure distribution of the installation. The increase of the tunnel section causes the reduction of the dynamic pressure, i.e. as we move away from the inlet the dynamic pressure decreases.

In Figure 5.11 it is remarkable the decrease of the stagnation pressure, originated by the deceleration of the flow inside the tunnel. In the entrance section the stagnation pressure is around 101.305 kPa, while in the exit section it is around 101.259 kPa.

The modulus of the longitudinal component (X) of the velocity is much higher when compared to the other components (Y and Z), throughout the Tunnel. Therefore, we decided to disregard the Y and Z components, so in this paper only the longitudinal velocity component is presented. In Figure 5.12 the longitudinal velocity component is presented.

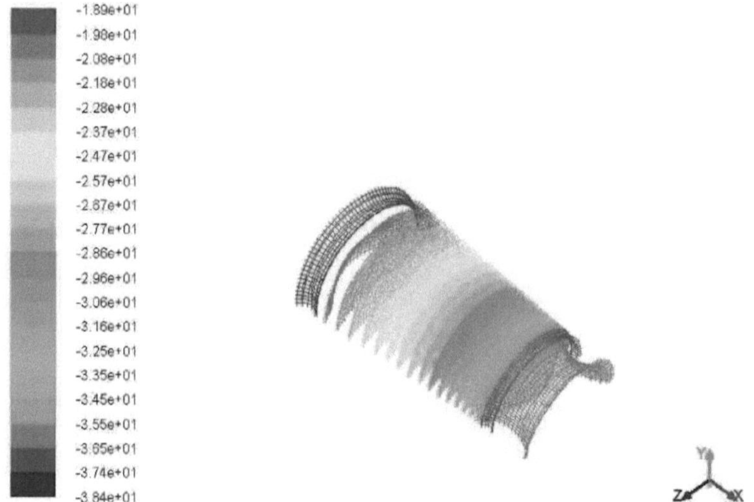

Figure 5.12 - Values of velocities in the longitudinal direction, for non-viscous flow, in [m/s].

Through Figure 5.12 it can be seen that the maximum module of the flow velocity occurs at the connection between element 1 and element 2 of the Tunnel and is around 37.4 m/s. The negative sign appears due to the fact that the flow occurs in the opposite direction to the X-axis.

For the same boundary conditions, simulations for viscous flow were also performed. Therefore, Figure 5.13 shows the static pressure distribution in the Tunnel for viscous flow with the k-c turbulence model.

Figure 5.13 - Static pressure distribution in the Tunnel, for viscous flow, in [Pa].

Figures 5.14 and 5.15 represent the dynamic and stagnation pressure distribution respectively, for viscous flow.

Figure 5.14 - Dynamic pressure distribution in the Tunnel, for viscous flow, in [Pa].

The maximum values of static pressures, presented in Figure 5.13, are limited between 101.004 kPa to 101.04 kPa. Whereas for the non-viscous flow, the maximum values of static pressures vary between 101.015 kPa and 101.047 kPa.

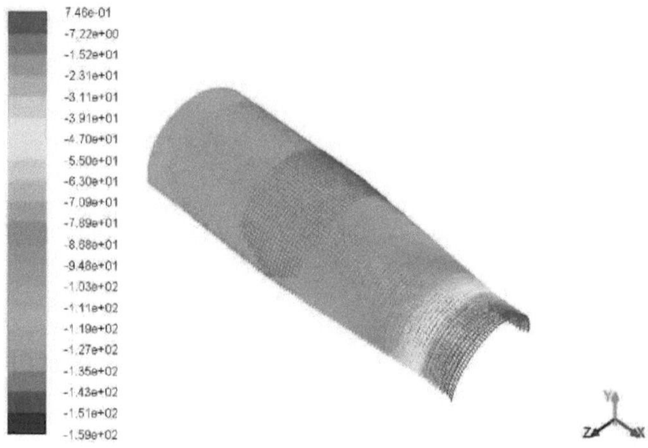

Figure 5.15 - Stagnation pressure distribution in the Tunnel, for viscous flow, in [Pa].

The stagnation pressure in the inlet section is around 101.293 kPa, and for the outlet section it is around 101.173 kPa.

In the non-viscous flow image, only the X-component of the velocity for viscous flow is also shown, and the other components have been neglected. Figure 5.16 shows the X-component of the velocity for viscous flow, with standard k-c turbulence model.

Figure 5.16 - Values of velocities in the longitudinal direction, for viscous flow, in [m/s].

The maximum module of the flow velocity, shown in Figure 5.16, occurs inside the stabilization stator and is approximately equal to 37 m/s. This value decreases in the flow direction. Again, the negative sign appears due to the fact that the flow occurs in the opposite direction to the X-axis.

To obtain better accuracy in the results we opted to refine the mesh, having increased the number of elements and nodes to **1,451,739** and **343,713**, respectively. As expected, this simulation had a longer running time (2h30 CPU) compared to the previous one (10 min CPU). Therefore, Figure 5.17 shows the velocity distribution of the flow.

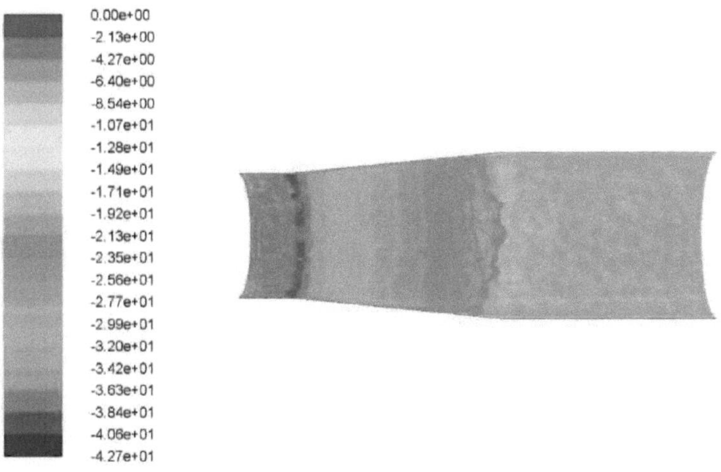

Figure 5.17 - Values of velocities in the longitudinal direction (simulation 3), for viscous flow, in [m/s].

By means of figure 5.17 it is possible to verify that the velocity of the inlet flow is around 38 m/s; this value corresponds to a deviation of 2.63% in relation to the value obtained during the previous computational simulation for viscous flow (37 m/s). The velocity profile inside the inlet section (blue part of the model) is typical of fully developed turbulent flow. More turbulence is seen inside the divergent element (green area), due to boundary layer separation. This turbulence is carried by convection to the outlet section. As expected, the velocity near the wall is zero (red zone of the model), due to the adherent conditions of the flow.

With the flow velocity values, the dynamic and stagnation pressure distributions are also presented in Figures 5.18 and 5.19, respectively.

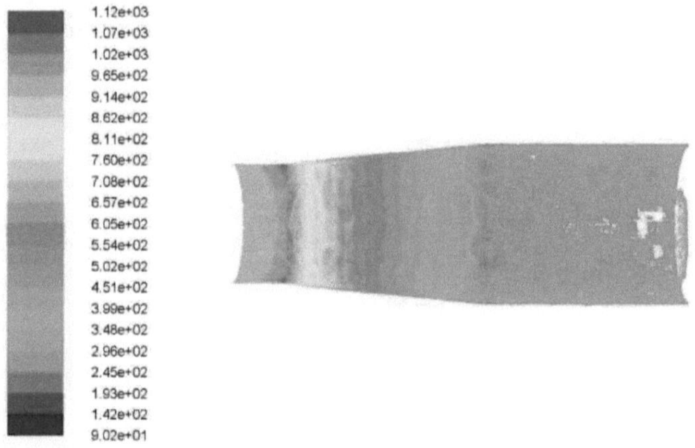

Figure 5.18 - Dynamic pressure distribution in the Tunnel (simulation 3), for viscous flow, in [Pa].

In the previous simulation for viscous flow, the dynamic inlet pressure is 822 Pa (Fig. 5.14). In this last simulation the value found is 965 Pa, which corresponds to a deviation of 14.8% in relation to the previous value. The increase of section in the divergent element provides a loss of energy in the flow direction, which is associated to the displacement of the boundary layer.

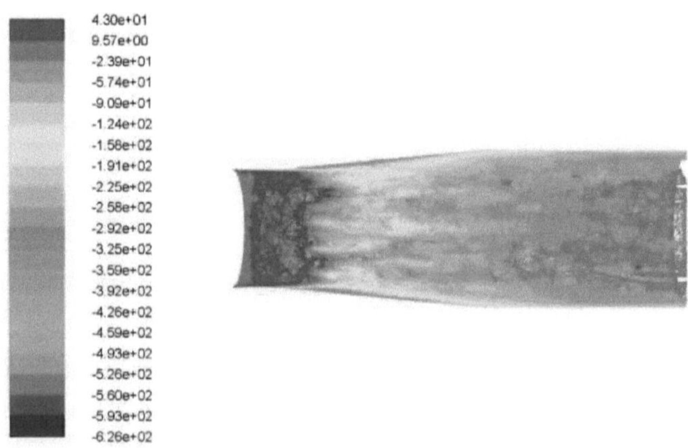

Figure 5.19 - Stagnation pressure distribution in the Tunnel (simulation 3), for viscous flow, in [Pa].

In Figure 5.19 it is observed that the boundary layer is poorly developed inside the smaller section. It develops substantially along the divergent element. It is also observed an intense diffusion of the quantity of movement, by viscous action, which culminates with the stabilization of the velocity profile in the final section of the tunnel, as can be seen in Figure 5.17.

The computational simulation for viscous and non-viscous flow allowed us to draw a parallel with the results obtained in the two tests performed. The values of the relative static pressure distribution obtained in the tests (Table 4.7) are close to the values obtained in the computer simulations (Figs. 5.9 and 5.13). As an example, the absolute static pressure obtained in the tests through outlet B is 100.723 kPa and, in the same section, the value obtained in the simulation for viscous flow is almost 100.427 kPa. This corresponds to a deviation of 0.294%. For non-viscous flow the value of the absolute static pressure obtained from the simulation is almost 100.428 kPa, which corresponds to a deviation of 0.293% from the experimental value.

Through Figures 5.9 and 5.13 it is possible to verify the presence of an adverse pressure gradient, characterized by pressure increase and velocity decrease in the opposite direction to the longitudinal axis (X). This gradient contributes to the decrease of the velocity of the fluid particles.

The energy losses occurring due to factors such as turbulence of the flow regime and possible leaks in the installation are quantified by the difference in stagnation pressure between the tunnel inlet and the tunnel outlet. For non-viscous flow, according to figure 5.11, the stagnation pressures at the inlet and outlet are 101.305 kPa and 101.260 kPa, respectively. Therefore, for non-viscous flow, the installation loss value is 45.7 Pa. Whereas for viscous flow (Fig. 5.15) the values of the stagnation pressures at the inlet and outlet are 101.293 kPa and 101.173 kPa, respectively, corresponding to 120 Pa of losses. As can be seen, the pressure losses in viscous flow are greater. This occurs due to the fact that in viscous flow there is presence of viscous stresses, responsible for originating a deformation rate to the fluid element, represented through the viscous term of the Navier-Stokes equation.

The velocity of the flow in the inlet section, obtained from the computer simulation, is approximately equal to 35.5 m/s for non-viscous flow (Fig. 5.12) and 37 m/s for viscous flow (Fig. 5.16). In tests 1 and 2 the velocities in this section are 16.860 m/s and 15.088 m/s respectively (tables 4.4 and 4.10). The difference between the values of the measured velocities in the inlet section (tables 4.4 and 4.10) and the values obtained from the simulations (Figs. 5.12 and 5.16) is considerable. One of the factors that contributed to this difference was the fact that during the measurements at the Tunnel entrance the probe was positioned almost on the outside of this section. Contrary to the software used, which considers exactly the tunnel entrance section.

For the outlet section, the velocities from the simulations performed are approximately equal to 19.8 m/s for the non-viscous flow (Fig. 5.12) and 14.13 m/s for the viscous flow (Fig. 5.16). In the same section, the velocities of the measurements were 10.584 m/s in test 1 (table 4.6) and 10.004 m/s in test 2 (table 4.12). The disparity between the test and simulation values is partly justified by the fact that the experimental measurements were taken downstream of the fan impeller. Contrary to the values of the computer simulations, where the rotor element is not considered.

The mass flow rates obtained in the two tests were compared to those obtained in the computer simulations. Table 5.4 shows the values of these mass flow rates and their respective origins.

Table 5.4: Mass flow rates.

Mass flow rate F?l	Simulation for non-viscous flow		Simulation for viscous flow		Test 1		Test 2	
	Enter	Output	Enter	Output	Enter	Output	Enter	Output
	6,170	6,170	6,343	6,343	5,319	5,161	5,204	5,464

For the inlet section, the mass flow rate obtained from test 1 shows a deviation of 13% from the mass flow rate obtained from the simulation for non-viscous flow and 16% from the computer simulation for viscous flow. While the mass flow rate at the outlet, in test 1, shows a deviation of 16% from the simulation value for non-viscous flow and 19% for viscous flow.

In test 2, the inlet flow rate has a deviation of 16% in relation to the simulation value for non-viscous flow and 18% in relation to viscous flow. For the outlet flow rates, this difference is 11% and 14% for non-viscous and viscous flow, respectively.

In general, computer simulations aim to validate the data obtained through the installation tests and to identify the values of pressures and velocities in areas of difficult access in the experimental environment. The various simulations performed served as support for the design and development of a wind tunnel for testing Turbomachinery, operating in the *fluidslab* laboratory of UBI.

Conclusion

The present work consisted of the assembly and testing of a low speed (37 m/s) circular cross section steel wind tunnel. The installation assembled in the *fluidslab* consists of a wind tunnel composed of an inlet section, a stabilization stator, a divergent nozzle and a cylindrical element of constant section, which is coupled to a 2.94 kW fan and respective discharge nozzle. The Tunnel operates by suction, and the Turbomachinery to be tested shall be coupled to the inlet section (section A). This way, there will be little disturbance of the flow in the test area.

The main body of the tunnel is supported on two steel structures designed and manufactured during the preparation of this work, with the aid of Solidworks® 2013 software. It is planned, in the future, to join these two structures by two steel beams. The design of another constant section nozzle is in progress, which will allow the addition of a second fan to the propulsion system of the Tunnel. Once this element is built and installed, we will be able to reposition and fix the support structures. On the other hand, the fabrication of the wooden support elements (Fig. 3.2) and the steel bars (Fig. 3.5) are part of a series of activities planned for the final finishing of the installation.

In order to improve the measurement and calibration systems of the installation, it is also underway the manufacture of a ring in MDF type wood, with a lateral opening that facilitates the entry and positioning of the Prandtl probe. This ring will be positioned between Tunnel element 1 and element 2, thus allowing a better characterisation of the flow in the inlet section. In practical terms this allows to correct the measurements previously performed, since they were made slightly outside the inlet section.

To obtain the essential characteristics of the facility were determined the Reynolds and Mach numbers in the entrance and exit sections of the Tunnel. The Mach number at the entrance varies between 0.044 and 0.049, and at the exit it is between 0.0268 and 0.027. The Reynolds number at the entrance varies between 6.03x105 and 6.74x105; for the exit section, the values are between 4.91x105 and 5.03x105. An important finding is that the flow in the Tunnel is turbulent developed.

The velocity of the flow in the entrance section of the Tunnel varies between 35,2 m/s and 37 m/s. The absolute static pressure in the zone between element 1 and 2 is 100.723 kPa, whereas in the zone between element 2 and 3 the pressure is 101.006 kPa. The energy losses, translated as stagnation pressure losses in the system, are on the order of 120 Pa (0.12% of the total inlet pressure). As can be seen, the energy loss along the nozzle is small, thanks to the low opening angle of the diffuser (5.6°).

Since we did not have a variable speed drive available, we did not test the installation for different fan speeds. This fact made it impossible to deduce the characteristic curves of the installation, as well as to present the speed profiles in different operating conditions of the fan.

After having performed the experimental test in the Wind Tunnel, we sought to validate the results obtained with the aid of Fluent® 2016 software. In this context, computer simulations were performed for non-viscous flow and viscous flow. For the viscous flow, the k-c turbulence model was used. Since the flow is aligned with the geometry of the model, meshes structured with quadrilateral elements were used. The conditions imposed on the boundaries of our computational model are those obtained from the experimental tests of the facility. The computational simulation performed allowed us to obtain the distributions of static pressure, dynamic pressure, stagnation pressure and velocity in several sections along the Tunnel. Through Figure 5.17 it was possible to visualise greater turbulence in the zone of the divergent element, caused by the boundary layer separation. This does not harm our installation, since all turbulence is dragged to the fan and expelled out of the Tunnel. This way we will have an inlet section with little disturbed flow.

The culmination of the simulations carried out and the calibration work results from the comparison between the mass flow rate values obtained in the computer simulation and those determined through the experimental tests. The flow rate of the installation is approximately equal to 5.464 kg/s. This value corresponds to the experimental flow rate that presents the least deviation (14%) from the flow rate value obtained through the simulation for viscous flow.

In general, our Wind Tunnel meets the conditions for testing air turbines in subsonic turbulent flow. Furthermore, the design and manufacture of several elements (already mentioned) is underway, which will improve the Facility's efficiency.

Bibliography

1. F. M. Roque (2011): Contribuição para o desenvolvimento de túnel de vento para o ensaio de coras de blades de Turbomáquinas, master's thesis, Universidade da Beira Interior, Covilhã.

2. S. Okomoto, Wind tunnels, ed. S.Croatia: Okomoto, 2011.

3. J. C. Lerner, U. Boldes, Wind Tunnels and experimental fluid dynamics research, Ed. Intech, Croatia, 2011.

4. J. D. Pereira, Wind tunnels - Areodynamics, models and experiments, Ed. Justin D. Pereira, New York, 2011.

5. M. A. Gonzalez, N. Ahmed, J. N. Libii, Y. Yokoi & A. M. Souza, Wind tunnel designs and their diverse engineering applications, Ed. Croatia: N. A. Ahmed, 2013.

6. J.B. Barlow, W. H. Rae & A. Pope, Low speed wind tunnel testing, [3rd] ed, 1999.

7. J. Green & J. Quest (2011): A short history of the European transonic wind tunnel ETW, Progress in Aerospace Sciences, pp. 319-368, Germany.

8. C. Meher: The Historical Evolulution of Turnomachinery, Proceeding of the 29thTurbochinery Symposium, p. 281 - 322, California.

9. J. Bretheim, E. Bardy (2012): A Review of Power-Generating Turbomachines, proceeding of the 2012 ASEE North Central Section Conference, p. 1 - 15, Pennsylvania.

10. N. Zawodny & H. Haskin (2017): Small propeller and rotor testing capabilities of the NASA Langley low speed aeroacoustic wind tunnel, American Institute of Aeronautics and Astronautics, p. 1 - 17, USA.

11. D. M. Elliott (2012): Initial investigation of the acoustics of a counter-rotating open rotor model with historical baseline blades in a low-speed wind tunnel, p. 1 - 22, USA.

12. J. Bell (2015): Advancing Test Capabilities at NASA Wind Tunnels, Presentation for the 32nd Annual International Test and Evaluation Symposium, p. 1 - 24.

13. M. Talavera, F. Shu (2017): Experimental study of turbulence intensity influence on wind turbine performance and wake recovery in a low-speed wind tunnel, Renewable Energy, p. 363 - 371, New Mexico State University, USA.

14. D. Corriveau & S. a. Sjolander (2002): Impact of Flow Quality in Transonic Cascade Wind Tunnels, ICAS Congress, p. 1 - 13, Ottawa, Canada.

15. K. David, J. Gorham, S. Kim, P. Miller & C. Minkus(2006): Aeronautical wind tunnels Europe and Asia, a report prepared by the Federal Research Division, Library of Congress, pp. 364, Washington. D. C, U.S.A.

16. R. Shaw, A. Lewkowicz& J. Gostelow(1966): Measurement of Turbulence in the Liverpool University Turbomachinery Wind Tunnels and Compressor, Aeronautical Research Council, p. 1 - 17, Liverpool. .

17. I.Hussain, M. Majeed, A. Ali & W. Sarsam (2011): Design, Construction and Testing of Low Speed Wind Tunnel with Measurement and Inspection Devices, Journal of Engineering, p.2 - 16, Baghdad.

18. I. Hussain & A. Ali (2014): Testing and Commissioning of a Low-Speed Wind Tunnel (LSWT) Test Section, Journal of Engineering, p. 106 - 125, Baghdad

19. I. Bayati, M. Belloli, L. Bernini & A. Zasso (2017): Aerodynamic design methodology for wind tunnel tests of wind turbine rotors, Journal of Wind Engineering and Industrial Aerodynamic, p. 217 - 227, Italy.

20. E.Zanoun (2017): Flow characteristics in low-speed wind tunnel contractions: simulation and testing, Alexandria engineering journal, p. 1 -13, Egypt.

21. F.M.White, Fluid Mechanics, 6th ed. New York, USA: McGraw-Hill bookman, 2007.

22. J. Tu, G. Yeoh &C.Liu, Computational Fluid Dynamics, 2nd ed.

List of websites consulted

A.1 http://www.grc.nasa.gov/WWW/K-12/WindTunnel/history.html

A.2 http://www.grc.nasa.gov/WWW/K-12/airplane/wrights/test1901.html

A.3 http://nyethermodynamics.com/primer/index.html

A.4 http://gasturbineworking.blogspot.com/2011/04/

A.5 https://www.sheffield.ac.uk/mecheng/staff/rhowell2

A.6 https://www.vki.ac.be/index.php/research-consulting-mainmenu-107/facilities-other
menu-148/low-speed-wt-other-menu-151/57-3meter-diameter-wind-tunnel-l-1

A.7 https://www.vki.ac.be/index.php/research-consulting-mainmenu-107/facilities-other
menu-148/low-speed-wt-other-menu-151/64-adaptive-wall-wind-tunnel-ta-3

A.8 http://www.vzlu.cz/en/a-wind-tunnel-for-turbo-machinery-purposes-
c74.html#prettyPhoto

A.9 https://www.chalmers.se/en/departments/m2/simulator-labs/labs/Pages/Chalmers-
windtunnels.aspx

A.10 https://www.padtinc.com/blog/wp-content/uploads/2017/04/Advanced-Techniques- in-
ANSYS-Meshing_Blog.pdf

A.11 https://www.sharcnet.ca/Software/Fluent6/html/ug/node478.htm

Annexes

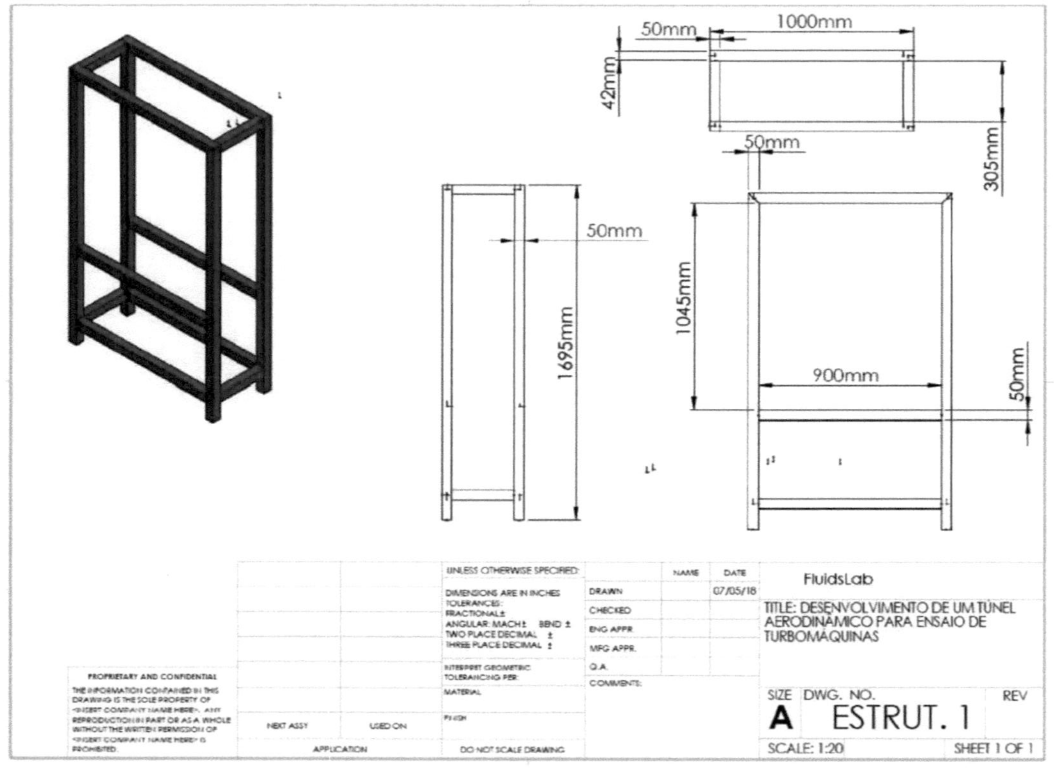

1000mm
50mm
42mm
305mm
50mm
1045mm
900mm
50mm
1695mm
50mm
UNLESS OTHERWISE SPECIFIED:
DIMENSIONS ARE IN INCHES
TOLERANCES:
FRACTIONAL±
ANGULAR: MACH±   BEND ±
TWO PLACE DECIMAL   ±
THREE PLACE DECIMAL   ±
INTERPRET GEOMETRIC
TOLERANCING PER:
MATERIAL
FINISH
NAME   DATE
DRAWN   07/05/18
CHECKED
ENG APPR.
MFG APPR.
Q.A.
COMMENTS:
FluidsLab
TITLE: DESENVOLVIMENTO DE UM TÚNEL AERODINÂMICO PARA ENSAIO DE TURBOMÁQUINAS
SIZE   DWG. NO.   REV
A   ESTRUT. 1
SCALE: 1:20   SHEET 1 OF 1
PROPRIETARY AND CONFIDENTIAL
THE INFORMATION CONTAINED IN THIS DRAWING IS THE SOLE PROPERTY OF <INSERT COMPANY NAME HERE>. ANY REPRODUCTION IN PART OR AS A WHOLE WITHOUT THE WRITTEN PERMISSION OF <INSERT COMPANY NAME HERE> IS PROHIBITED.
NEXT ASSY   USED ON
APPLICATION
DO NOT SCALE DRAWING

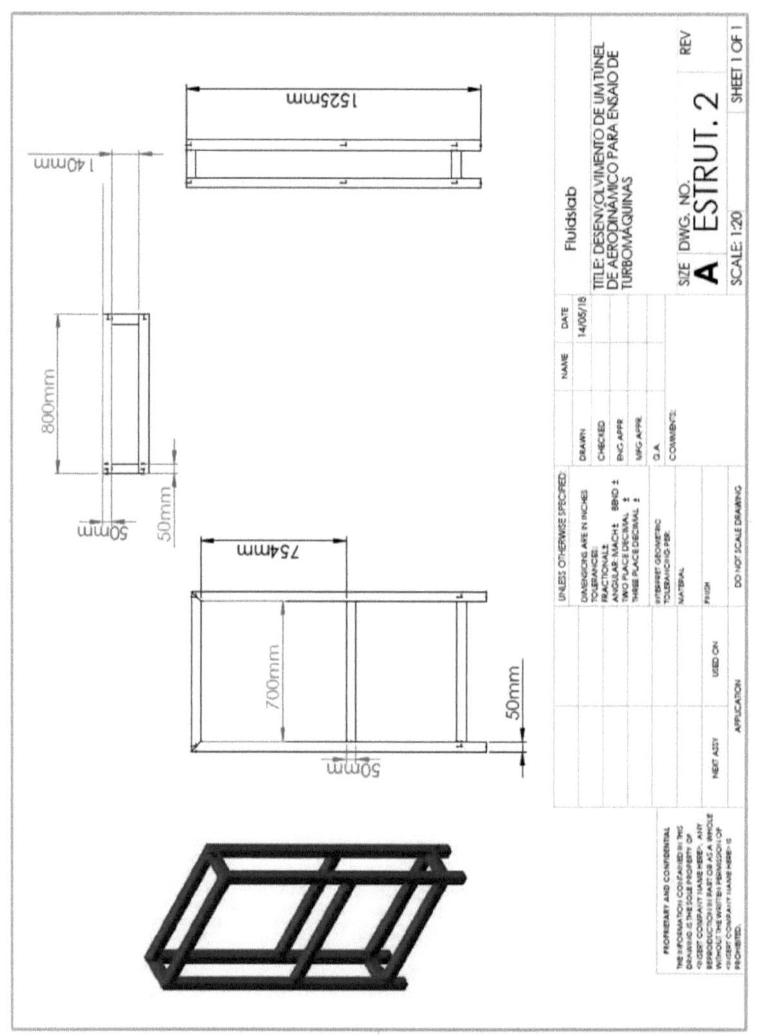

1525mm
140mm
800mm
50mm
50mm
754mm
700mm
50mm
50mm
Fluidslab
DATE
NAME
14/05/18
UNLESS OTHERWISE SPECIFIED
DIMENSIONS ARE IN INCHES
TOLERANCES:
FRACTIONAL±
ANGULAR: MACH±   BEND ±
TWO PLACE DECIMAL ±
THREE PLACE DECIMAL ±
INTERPRET GEOMETRIC
TOLERANCING PER:
MATERIAL
FINISH
DRAWN
CHECKED
ENG APPR
MFG APPR
Q.A.
COMMENTS:
TITLE: DESENVOLVIMENTO DE UM TÚNEL
DE AERODINÂMICO PARA ENSAIO DE
TURBOMÁQUINAS
SIZE DWG. NO.   REV
A ESTRUT. 2
SCALE: 1:20   SHEET 1 OF 1
DO NOT SCALE DRAWING
NEXT ASSY   USED ON
APPLICATION
PROPRIETARY AND CONFIDENTIAL
THE INFORMATION CONTAINED IN THIS
DRAWING IS THE SOLE PROPERTY OF
<INSERT COMPANY NAME HERE>. ANY
REPRODUCTION IN PART OR AS A WHOLE
WITHOUT THE WRITTEN PERMISSION OF
<INSERT COMPANY NAME HERE> IS
PROHIBITED.

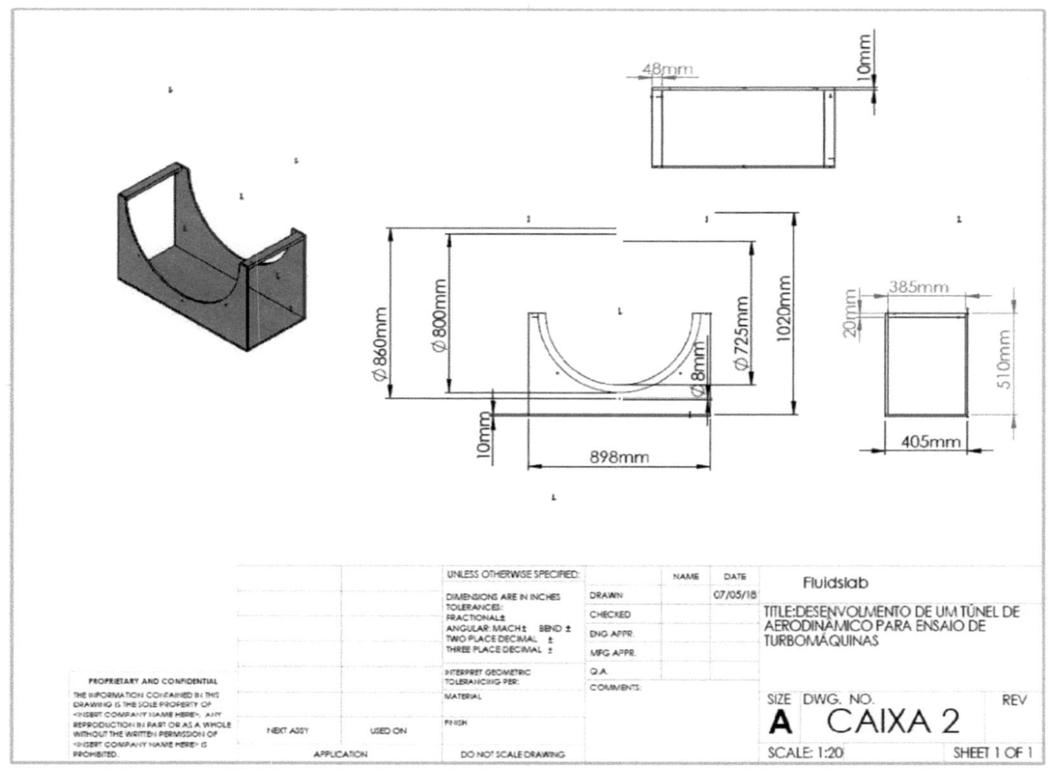
48mm
10mm
Ø 860mm
Ø 800mm
Ø 725mm
8mm
1020mm
10mm
898mm
385mm
20mm
510mm
405mm
UNLESS OTHERWISE SPECIFIED:
DIMENSIONS ARE IN INCHES
TOLERANCES:
FRACTIONAL±
ANGULAR: MACH±   BEND ±
TWO PLACE DECIMAL   ±
THREE PLACE DECIMAL ±
INTERPRET GEOMETRIC
TOLERANCING PER:
MATERIAL
FINISH
NAME
DATE
DRAWN
CHECKED
ENG APPR.
MFG APPR.
Q.A.
COMMENTS:
07/05/18
Fluidslab
TITLE:DESENVOLVIMENTO DE UM TÚNEL DE
AERODINÂMICO PARA ENSAIO DE
TURBOMÁQUINAS
SIZE DWG. NO.
A   CAIXA 2
REV
SCALE: 1:20
SHEET 1 OF 1
PROPRIETARY AND CONFIDENTIAL
THE INFORMATION CONTAINED IN THIS
DRAWING IS THE SOLE PROPERTY OF
<INSERT COMPANY NAME HERE>. ANY
REPRODUCTION IN PART OR AS A WHOLE
WITHOUT THE WRITTEN PERMISSION OF
<INSERT COMPANY NAME HERE> IS
PROHIBITED.
NEXT ASSY
USED ON
APPLICATION
DO NOT SCALE DRAWING

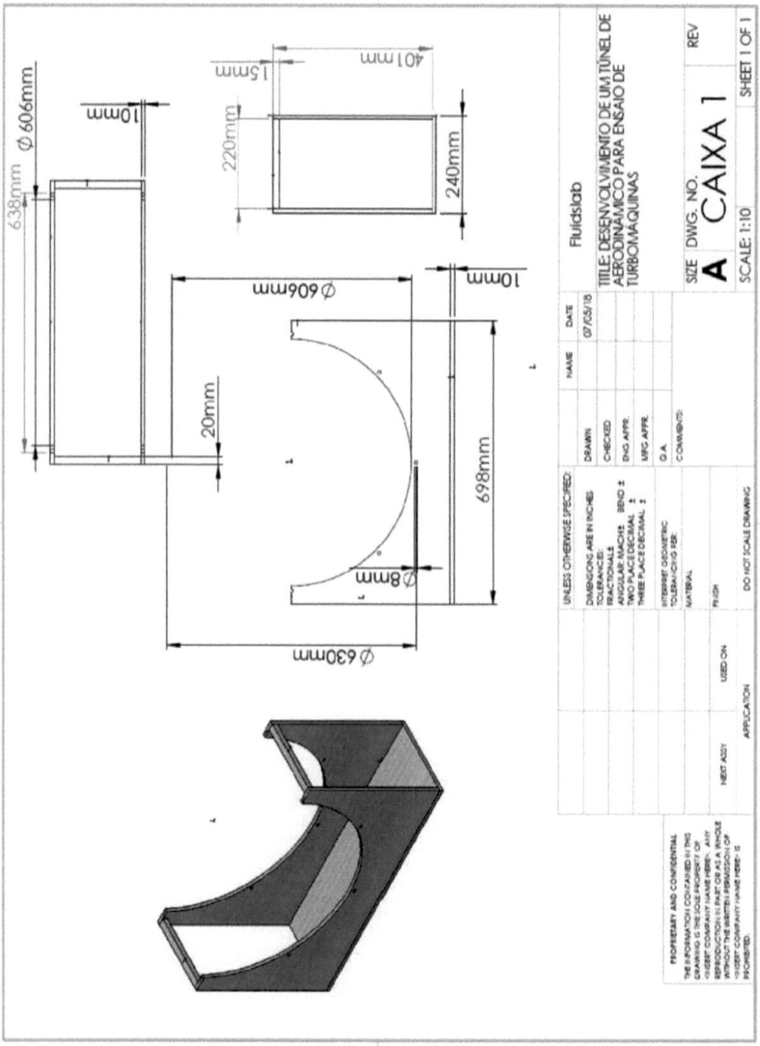

UNLESS OTHERWISE SPECIFIED:		NAME	DATE	Fluidslab		
DIMENSIONS ARE IN INCHES TOLERANCES: FRACTIONAL± ANGULAR: MACH± BEND ± TWO PLACE DECIMAL ± THREE PLACE DECIMAL ±	DRAWN		07/05/18	TITLE: DESENVOLVIMENTO DE UM TÚNEL DE AERODINÂMICO PARA ENSAIO DE TURBOMÁQUINAS		
	CHECKED					
	ENG APPR.					
	MFG APPR.					
INTERPRET GEOMETRIC TOLERANCING PER:	Q.A.					
MATERIAL	COMMENTS:			SIZE **A**	DWG. NO. CAIXA 1	REV
FINISH				SCALE: 1:10		SHEET 1 OF 1

PROPRIETARY AND CONFIDENTIAL

THE INFORMATION CONTAINED IN THIS DRAWING IS THE SOLE PROPERTY OF <INSERT COMPANY NAME HERE>. ANY REPRODUCTION IN PART OR AS A WHOLE WITHOUT THE WRITTEN PERMISSION OF <INSERT COMPANY NAME HERE> IS PROHIBITED.

NEXT ASSY   USED ON

APPLICATION

DO NOT SCALE DRAWING

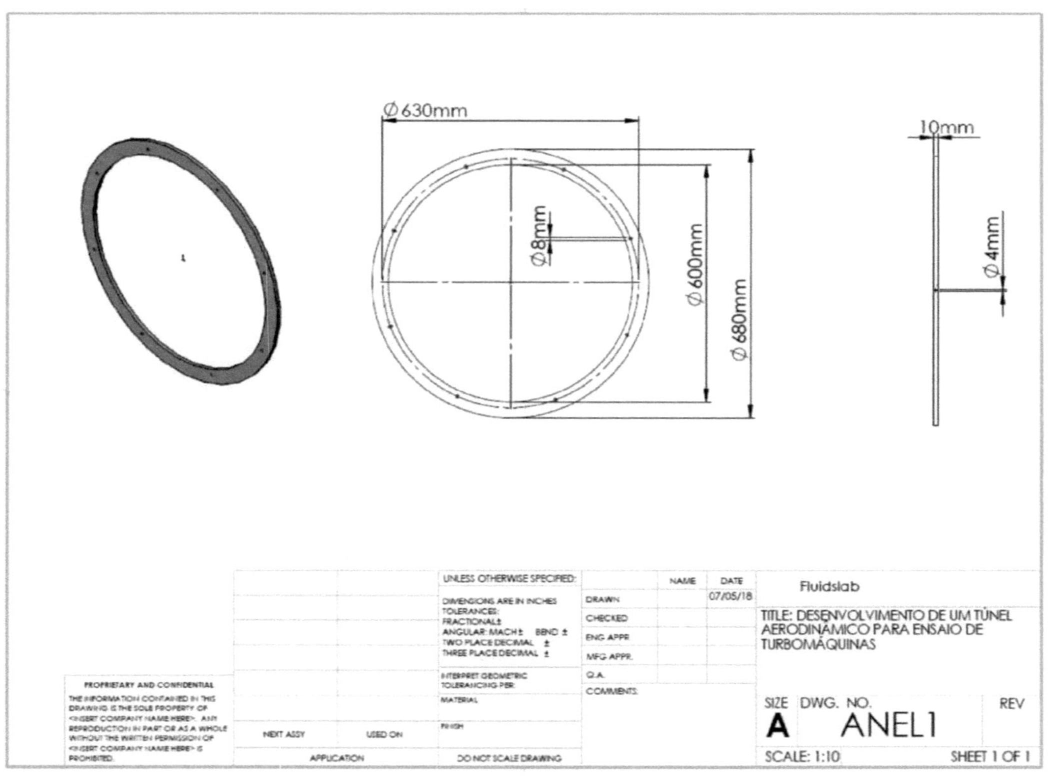

Ø 630mm
Ø 8mm
Ø 600mm
Ø 680mm
10mm
Ø 4mm
UNLESS OTHERWISE SPECIFIED:
DIMENSIONS ARE IN INCHES
TOLERANCES:
FRACTIONAL±
ANGULAR: MACH ±   BEND ±
TWO PLACE DECIMAL   ±
THREE PLACE DECIMAL ±
INTERPRET GEOMETRIC
TOLERANCING PER:
MATERIAL
FINISH
NAME   DATE
DRAWN   07/05/18
CHECKED
ENG APPR
MFG APPR.
Q.A.
COMMENTS:
Fluidslab
TITLE: DESENVOLVIMENTO DE UM TÚNEL
AERODINÂMICO PARA ENSAIO DE
TURBOMÁQUINAS
SIZE DWG. NO.   REV
A   ANEL1
SCALE: 1:10   SHEET 1 OF 1
PROPRIETARY AND CONFIDENTIAL
THE INFORMATION CONTAINED IN THIS
DRAWING IS THE SOLE PROPERTY OF
<INSERT COMPANY NAME HERE>. ANY
REPRODUCTION IN PART OR AS A WHOLE
WITHOUT THE WRITTEN PERMISSION OF
<INSERT COMPANY NAME HERE> IS
PROHIBITED.
NEXT ASSY   USED ON
APPLICATION   DO NOT SCALE DRAWING

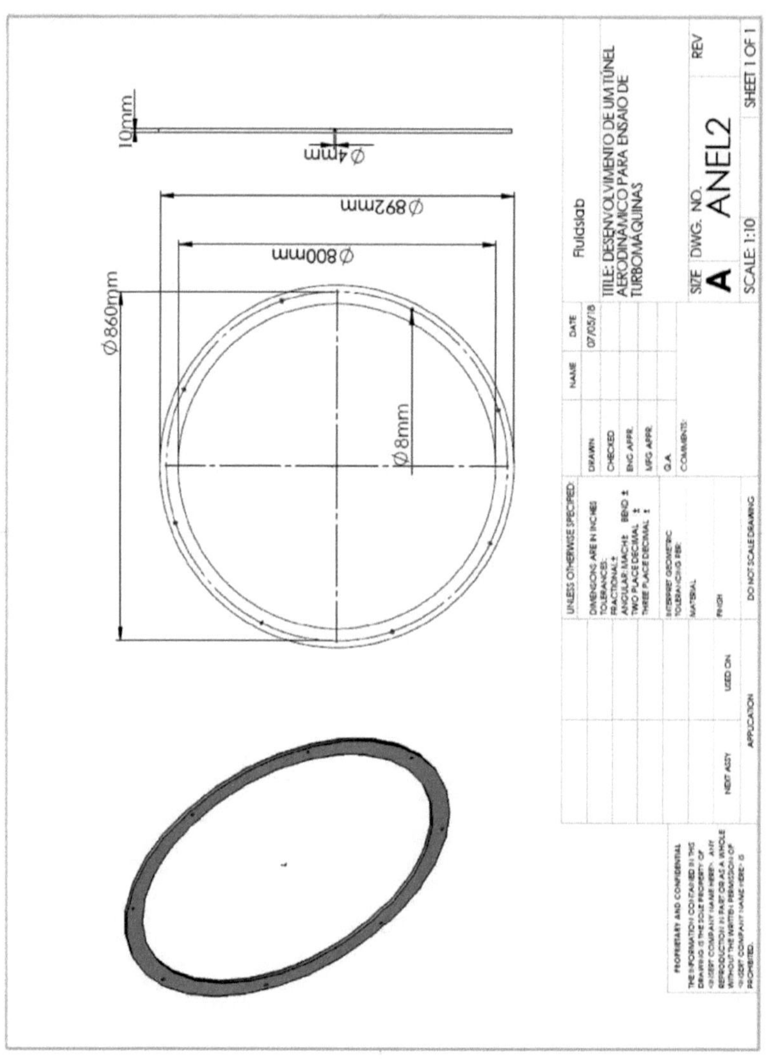

10mm
Ø4mm
Ø892mm
Ø800mm
Ø860mm
Ø8mm
Fluidslab
TITLE: DESENVOLVIMENTO DE UM TÚNEL AERODINÂMICO PARA ENSAIO DE TURBOMÁQUINAS
SIZE A
DWG. NO. ANEL2
REV
SCALE 1:10
SHEET 1 OF 1
UNLESS OTHERWISE SPECIFIED:
DIMENSIONS ARE IN INCHES
TOLERANCES:
FRACTIONAL
ANGULAR: MACH± BEND ±
TWO PLACE DECIMAL ±
THREE PLACE DECIMAL ±
INTERPRET GEOMETRIC TOLERANCING PER:
MATERIAL
FINISH
DO NOT SCALE DRAWING
NAME DATE
DRAWN 07/05/18
CHECKED
ENG APPR.
MFG APPR.
Q.A.
COMMENTS:
NEXT ASSY   USED ON
APPLICATION
PROPRIETARY AND CONFIDENTIAL
THE INFORMATION CONTAINED IN THIS DRAWING IS THE SOLE PROPERTY OF <INSERT COMPANY NAME HERE>. ANY REPRODUCTION IN PART OR AS A WHOLE WITHOUT THE WRITTEN PERMISSION OF <INSERT COMPANY NAME HERE> IS PROHIBITED.

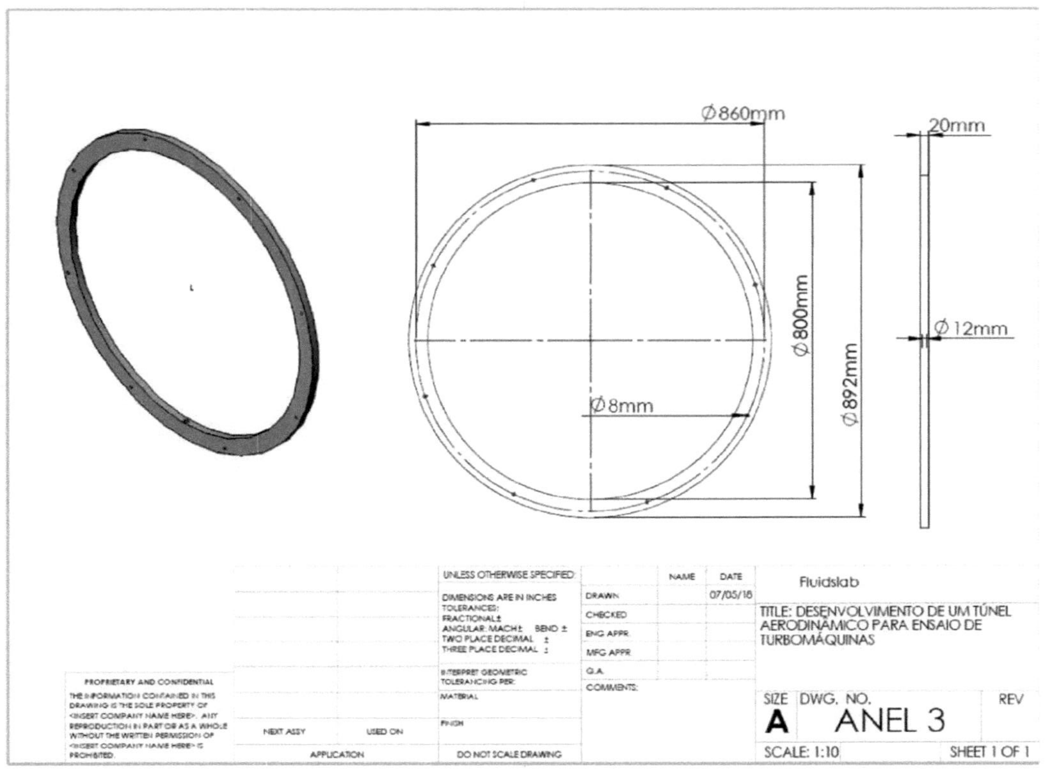

Ø860mm
20mm
Ø800mm
Ø12mm
Ø892mm
Ø8mm
UNLESS OTHERWISE SPECIFIED:
DIMENSIONS ARE IN INCHES
TOLERANCES:
FRACTIONAL±
ANGULAR: MACH± BEND ±
TWO PLACE DECIMAL ±
THREE PLACE DECIMAL ±
INTERPRET GEOMETRIC TOLERANCING PER:
MATERIAL
FINISH
NAME
DATE
DRAWN
07/05/18
CHECKED
ENG APPR
MFG APPR
Q.A.
COMMENTS:
Fluidslab
TITLE: DESENVOLVIMENTO DE UM TÚNEL AERODINÂMICO PARA ENSAIO DE TURBOMÁQUINAS
SIZE
A
DWG. NO.
ANEL 3
REV
SCALE: 1:10
SHEET 1 OF 1
PROPRIETARY AND CONFIDENTIAL
THE INFORMATION CONTAINED IN THIS DRAWING IS THE SOLE PROPERTY OF <INSERT COMPANY NAME HERE>. ANY REPRODUCTION IN PART OR AS A WHOLE WITHOUT THE WRITTEN PERMISSION OF <INSERT COMPANY NAME HERE> IS PROHIBITED.
NEXT ASSY
USED ON
APPLICATION
DO NOT SCALE DRAWING

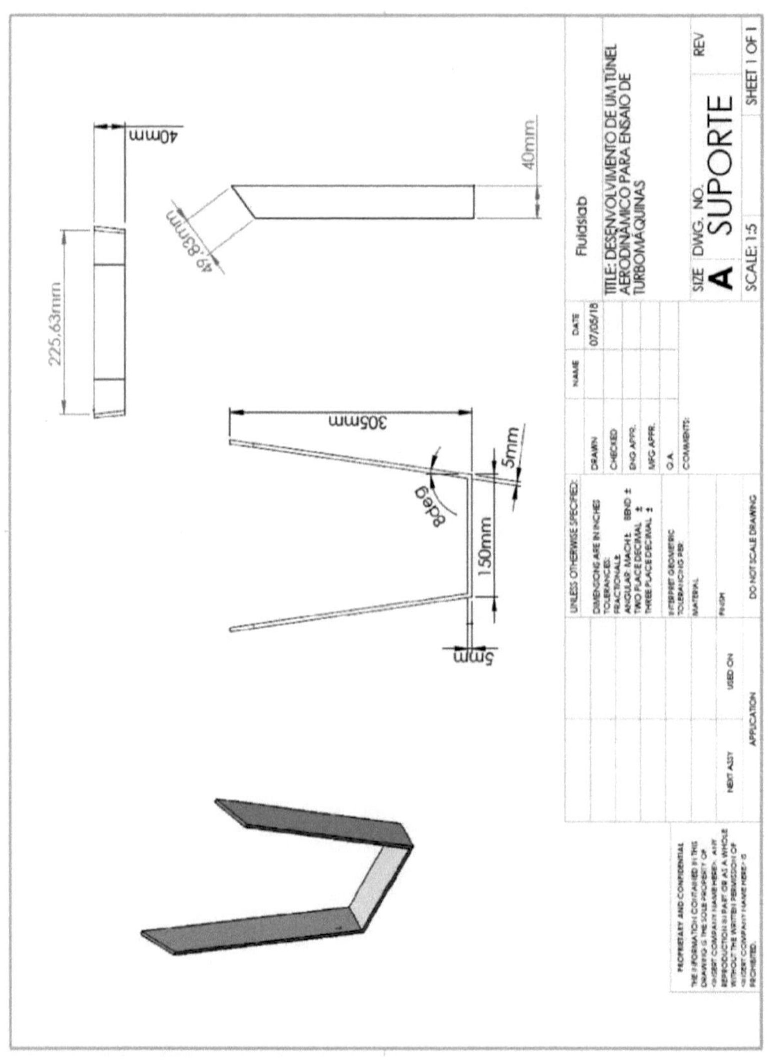

40mm
225,63mm
49,83mm
305mm
8deg
5mm
150mm
5mm
40mm
Fluidslab
TITLE: DESENVOLVIMENTO DE UM TUNEL AERODINÂMICO PARA ENSAIO DE TURBOMÁQUINAS
UNLESS OTHERWISE SPECIFIED:
DIMENSIONS ARE IN INCHES
TOLERANCES:
FRACTIONAL
ANGULAR: MACH ± BEND ±
TWO PLACE DECIMAL ±
THREE PLACE DECIMAL ±
INTERPRET GEOMETRIC TOLERANCING PER:
MATERIAL
FINISH
DO NOT SCALE DRAWING
NAME DATE
DRAWN 07/05/18
CHECKED
ENG APPR.
MFG APPR.
Q.A.
COMMENTS:
SIZE A DWG. NO. SUPORTE REV
SCALE: 1:5
SHEET 1 OF 1
NEXT ASSY USED ON
APPLICATION
PROPRIETARY AND CONFIDENTIAL
THE INFORMATION CONTAINED IN THIS DRAWING IS THE SOLE PROPERTY OF <INSERT COMPANY NAME HERE>. ANY REPRODUCTION IN PART OR AS A WHOLE WITHOUT THE WRITTEN PERMISSION OF <INSERT COMPANY NAME HERE> IS PROHIBITED.

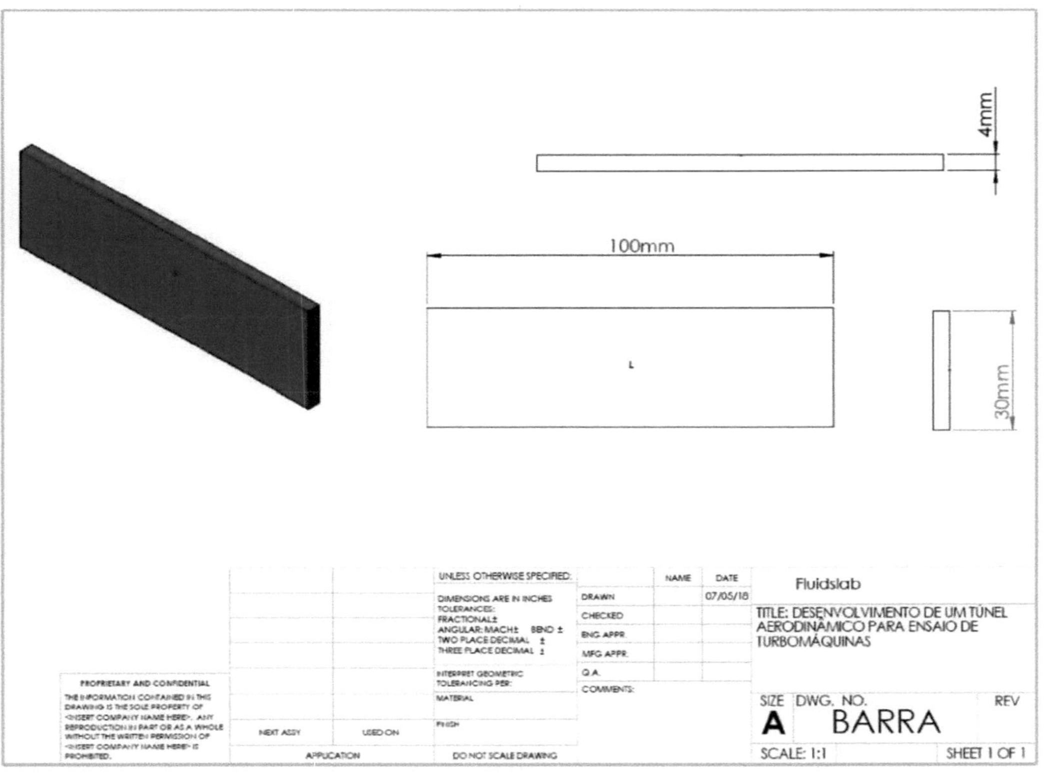

4mm
100mm
L
30mm
UNLESS OTHERWISE SPECIFIED:
DIMENSIONS ARE IN INCHES
TOLERANCES:
FRACTIONAL±
ANGULAR: MACH±   BEND ±
TWO PLACE DECIMAL   ±
THREE PLACE DECIMAL ±
INTERPRET GEOMETRIC
TOLERANCING PER:
MATERIAL
FINISH
NAME
DATE
DRAWN
07/05/18
CHECKED
ENG APPR.
MFG APPR.
Q.A.
COMMENTS:
Fluidslab
TITLE: DESENVOLVIMENTO DE UM TÚNEL AERODINÂMICO PARA ENSAIO DE TURBOMÁQUINAS
SIZE
A
DWG. NO.
BARRA
REV
SCALE: 1:1
SHEET 1 OF 1
PROPRIETARY AND CONFIDENTIAL
THE INFORMATION CONTAINED IN THIS DRAWING IS THE SOLE PROPERTY OF <INSERT COMPANY NAME HERE>. ANY REPRODUCTION IN PART OR AS A WHOLE WITHOUT THE WRITTEN PERMISSION OF <INSERT COMPANY NAME HERE> IS PROHIBITED.
NEXT ASSY
USED ON
APPLICATION
DO NOT SCALE DRAWING

Printed by Books on Demand GmbH, Norderstedt / Germany